Engineering Materials and Processes

T0180798

For further volumes:
http://www.springer.com/series/4604

Anastasios P. Vassilopoulos
Thomas Keller

Fatigue of Fiber-reinforced Composites

 Springer

Dr. Anastasios P. Vassilopoulos
Composite Construction Laboratory
 (CCLab)
École Polytechnique Fédérale de
 Lausanne (EPFL)
School of Architecture, Civil and
 Environmental Engineering (ENAC)
Station 16
1015 Lausanne
Switzerland
e-mail: anastasios.vasilopoulos@epfl.ch

Prof. Thomas Keller
Composite Construction Laboratory
 (CCLab)
École Polytechnique Fédérale de
 Lausanne (EPFL)
School of Architecture, Civil and
 Environmental Engineering (ENAC)
Station 16
1015 Lausanne
Switzerland
e-mail: thomas.keller@epfl.ch

ISSN 1619-0181
ISBN 978-1-4471-2694-2   ISBN 978-1-84996-181-3 (eBook)
DOI 10.1007/978-1-84996-181-3
Springer London Dordrecht Heidelberg New York

British Library Cataloguing in Publication Data
A catalogue record for this book is available from the British Library

Cover design: eStudio Calamar, Berlin/Figueres

Printed on acid-free paper

Springer is part of Springer Science+Business Media (www.springer.com)

Preface

Examples of composite materials abound in nature—wood, flower stems and bones are typical natural composite structural elements that have "adopted" the composite concept in order to develop and serve specific needs. Human history is also full of paradigms of the use of composite systems for particular applications—straw mixed with mud to fabricate bricks (Bible, Exodus 5.15-18) and laminated composite shields together with the description of weapon behavior in battle were presented in detail in the Homeric epics (Iliad 18.474-482).

It was not until the 20th century, however, that fiber-reinforced polymer composite materials were used as critical elements in emerging applications for which failure can mean anything from interrupted operation to catastrophic collapse with subsequent fatal accidents involving operators and users. Numerous structures of this kind can be found in the aerospace sector—significant parts of commercial and fighter airplanes are made of advanced composite materials; the automotive industry—racing car chassis, ceramic brakes, etc; the wind industry, where wind turbine rotors with diameters of more than 100 m are now constructed using fiber-reinforced composite materials and the civil engineering domain, with a substantial number of foot and vehicular bridges being constructed and numerous conceptual composite house designs.

All the above-mentioned structures encounter different loading profiles during their operational life and therefore must be designed to sustain all conceivable loading conditions, from ultimate static loads to time-dependent loads. Wind turbine rotor blades will probably be required to sustain 10^9 fatigue cycles during their 25 years of expected operational life, while a normal short-span vehicular bridge can sustain 10^7–10^8 fatigue cycles over a period of 50–70 years. Although most of these cycles are of low amplitude, they cause damage that is accumulated in the material and degrades its properties.

Anisotropy is one of the characteristics of composite materials. Depending on the degree of its anisotropy, a composite material can exhibit significantly different strength along different directions. Therefore, even loads of low amplitude compared to the strength of the stronger material direction can cause significant fatigue damage to the material under the development of a multiaxial stress field.

It is well documented that the majority of structural failures of composite and conventional structures occur due to fatigue and fatigue-related mechanisms. This percentage may be even higher if only composite structures are considered since for these lightweight structures live loads become critical in relation to the dead loads. For example a concrete road bridge is normally not fatigue-sensitive since the dead loads are considerably higher than the live loads, whereas a lightweight bridge is susceptible to fatigue (live) loads that constitute a large part of the total loading itself. Therefore, the fatigue of composite structures cannot be neglected and must be seriously taken into account in design processes in order to yield durable structures.

This book is intended as a guide for university students and new researchers to the world of the fatigue of fiber-reinforced composite materials. It will provide information regarding the fatigue behavior of the examined materials and lead the reader step by step through the available methods for the modeling of the fatigue life of fiber-reinforced composite materials and the prediction of their lifetime in the presence of complex, variable amplitude stress fields.

The analysis of the fatigue behavior exhibited by fiber-reinforced polymer composites and the presentation of the methods for their fatigue life modeling and prediction are based on an extensive database comprised of quasi-static, constant amplitude and variable amplitude fatigue experiments on on-axis and off-axis specimens cut from a multidirectional laminate with a stacking sequence composed of layers with fibers along the $0°$ and $±45°$ directions. The behavior of this material system, which is a common structural component for a number of engineering applications, under Tension–Tension, Tension–Compression and Compression–Compression constant amplitude fatigue loading patterns is described in detail in Chap. 2. A total of 324 fatigue experiments were performed under different fatigue loading conditions and 140.86 million fatigue cycles were recorded at a frequency of 10 Hz. The experimental results, together with results from the literature, were used to demonstrate the application of existing methodologies for the life modeling and life prediction of the examined materials.

The collected fatigue data are statistically analyzed in Chap. 3, where some of the most commonly used methods for the statistical analysis of constant amplitude fatigue data are presented. Modeling of the fatigue life based on the stress-life results and stiffness degradation measurements is demonstrated in Chap. 4 together with the presentation of the concept of the constant life diagrams that allow the estimation of S–N curves under unseen fatigue loading based on limited experimental input.

The application of the above-mentioned theory to a structural element, adhesively-bonded FRP joints, is demonstrated in Chap. 5 of this volume. The fatigue behavior of adhesively-bonded pultruded FRP joints under different temperatures is described, and it is shown how the developed and established S–N curve formulations can be applied for modeling the fatigue life of the examined joints under different temperatures.

Multiaxial fatigue failure criteria for estimating the fatigue life of a material system in the presence of complex plane stress states are presented in Chap. 6.

These criteria can take into account the synergistic effect of all stress tensor components on the strength and fatigue life of the examined material. A comparison of the predictive ability of the examined fatigue theories is also presented based on data from Chap. 2, and other data found in the literature.

Finally, Chap. 7 summarizes the knowledge accumulated in the previous chapters in order to demonstrate a simple methodology for the life prediction of composite materials under complex irregular stress states. The classic fatigue design methodology that is analytically presented is applied to the available variable amplitude fatigue data from Chap. 2 in order to demonstrate its applicability to the selected cases.

The writing of this book was a laborious task that took almost 3 years to complete and most of the results put forward have been published by the authors in international journals and presented at international scientific conferences. During this period the authors had the chance to receive valuable reviews and communicate with remarkable colleagues, experts in the field of composite material fatigue, in order to improve the content of this volume. Their assistance is acknowledged with gratitude.

The authors would also like to acknowledge the efforts of Mrs. Maria Lazari for the conversion of the numerous data into meaningful line-drawings, Ms. Margaret Howett for her professional copy-editing, and Mr. Behzad Dehghan Manshadi, who went through this volume from a critical point of view and contributed to the minimization of redundancies and conflicts between the content of the various chapters.

Anastasios P. Vassilopoulos
Thomas Keller

Basic Fatigue Nomenclature

The plethora of symbols and abbreviations used in different handbooks and other scientific publications makes difficult the derivation of a complete list of symbols referring to all the variables and parameters that encounter in the fatigue life modeling and prediction of FRP composite materials. Consistent representations of the different variables and parameters are used throughout this volume as much as possible, and defined when they appear. A list, containing the basic symbols and abbreviations is given below as well:

Table 1 List of symbols and abbreviations

σ	Stress variable
σ_a	Stress amplitude
σ_m	Mean cyclic stress
σ_{min}	Minimum cyclic stress
σ_{max}	Maximum cyclic stress
$\Delta\sigma$	Stress range
σ_e	Equivalent static strength (wear-out model, Eq. 3.26)
σ_r	Residual strength (wear-out model)
$R = \sigma_{min}/\sigma_{max}$	Stress ratio
ε	Strain
N	Number of cycles
N_f	Number of cycles to failure
fr	Fatigue frequency
UTS	Ultimate tensile stress
UCS	Ultimate compressive stress
X	Longitudinal tensile strength
X'	Longitudinal compressive strength
Y	Transverse tensile strength
Y'	Transverse compressive strength
S	In-plane shear strength
$F_{ij}, i,j = 1,2,6$	Failure tensor components
θ	Off-axis angle
σ_{st}	Off-axis static strength
$E(1)$	Young's modulus measured at the first cycle
$E(N)$	Young's modulus measured at the Nth cycle
$\sigma_o, k, A, B, \beta, C, G$	S–N curve formulation-model parameters

The basic fatigue terminology together with representative constant amplitude and variable amplitude loading patterns are schematically presented in Figs. 1, 2, and 3.

Fig. 1 Basic fatigue terminology

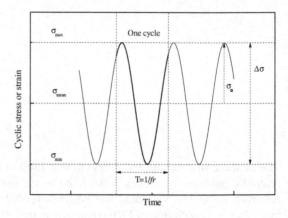

Fig. 2 Representative constant amplitude loading patterns

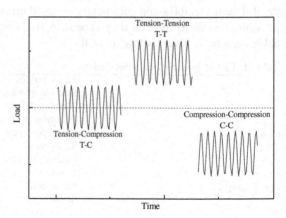

Fig. 3 Example of a variable amplitude fatigue time series

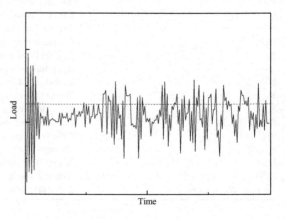

Contents

Chapter 1
Introduction to the Fatigue of Fiber-Reinforced Polymer Composites

1.1 Introduction

Composite materials offer numerous advantages compared to conventional isotropic media and the concept of composite structures was thus adopted very early
on in human history in order to create tools—mainly weapons—with enhanced
properties. Back in time we can trace for example the Mongolian horn bow arcs
that were made by mixing different materials. The compressed parts were made of
corn and the stressed parts were made of wood and cow tendons glued together to
obtain extraordinary strength.

References to highly interesting structures exhibiting elements of almost
modern technology can also be found in the Homeric epics. The shields of Achilles
and Ajax Telamonius—son of Telamon, king of Salamis—are presented as laminated structures consisting of successive layers of different metals and metal and
leather layers respectively and are characteristic examples of the advanced
knowledge of the science and technology of materials and structures possessed by
the Mycenaean Greeks [1]. Metals of particular interest in this respect were hard
bronze, tin and pure gold (shield of Achilles) and hard bronze and calf's leather
(shield of Ajax). The manufacture of Achilles' armor—by Hephaestus—is
described in detail in the Iliad. The description refers to a laminated composite
structure, consisting of five consecutive metal laminates with very different
mechanical properties. In fact, the shield consists of two external laminates of hard
bronze, two internal ones of tin and a central one of pure (soft) gold. This structure
exhibits maximum penetration resistance, as proved by a complete numerical
simulation of its elastoplastic behavior at large deformations, when impacted by
the tip of a piercing element, an arrow or spear for example, [1]. The shield of
Ajax presented a similar configuration, which according to Homer was a multi-
layered structure consisting of eight consecutive laminates—an external laminate
of hard bronze and seven layers of calf's leather underneath. The impressive
impact properties of this structure are also praised in the Iliad: "…The spear struck

A. P. Vassilopoulos and T. Keller, *Fatigue of Fiber-reinforced Composites*,
Engineering Materials and Processes, DOI: 10.1007/978-1-84996-181-3_1,
© Springer-Verlag London Limited 2011

the sevenfold shield in its outermost layer—the eighth which was of bronze—and went through six other layers but in the seventh hide it stayed..." from Iliad's description of the duel between Ajax and Hector. These unique detailed descriptions, which also include the weapon's battle behavior, constitute the first known applications of laminated structures in human history.

Composite structures were made based on the idea of mixing soft matrices with strong reinforcements, a concept copied from nature, with wood and bone being possibly the most common and easily comprehensible examples of natural composites. Wood consists of cellulose fibers, which give it the ability to bend without breaking, embedded in a compound called lignin, which provides the stiffening component. Bone is a combination of a soft form of protein known as collagen and a strong but brittle mineral called apatite. This concept has been adopted by humans for building applications (e.g., mixing mud and straw to make bricks) since very early times, or even for contemporary reinforced concrete, but it is only over the last century that composite materials have been used for advanced engineering structures in all fields ranging from aerospace (where the need for composites was determined by operational conditions) to automotive, civil and mechanical engineering in order to optimize existing designs and create novel more effective (light and durable) products.

Contemporary composite structures follow the same concept as those adopted 3,000 years ago, since they are constructed by combining a number of composite structural components in order to achieve improved properties and overall structural behavior. Nevertheless the term "advanced engineering composite material" has a somewhat different meaning today as it refers to the combination of a soft (matrix) and a strong (reinforcement) constituent that are combined in order to improve the properties of the matrix, facilitate the production process or assist load transfer between the stiff and strong reinforcements. Different types of matrices exist as well as different types of reinforcements. When the reinforcement is a fiber and the matrix is a polymeric resin the resulting material is known as a fiber-reinforced polymer or fiber-reinforced plastic (FRP) and, depending on the type of fiber (glass, carbon etc.), is designated glass fiber-reinforced polymer (GFRP), carbon fiber-reinforced polymer (CFRP) etc.

The characteristic of FRPs is that they are anisotropic and inhomogeneous in general and their mechanical behavior is very different from the mechanical behavior of metals and other isotropic materials. The failure of FRPs is not dominated by a single crack and its propagation as is common with metals but is rather a combination of successive phenomena (e.g., delamination, matrix failure, fiber pull-out and fiber failure) that act synergistically to accumulate damage in the material. FRP materials encounter different loading patterns during their theoretically long operational life. On the other hand, extensive studies [2] have shown that fatigue and related phenomena are the most frequent causes of structural failures in engineering structures, and it is widely accepted that fatigue must never be disregarded in any design process.

Fig. 1.1 The Eyecatcher
building in Basel Switzerland
(1998-source www.cclab.ch)

1.2 Fiber-Reinforced Composites in Engineering Applications

FRPs are used today instead of such homogeneous isotropic materials as steel, concrete and even the anisotropic wood that has long been employed in numerous applications. However, the engineer's perception was limited to the use of composites to replace the conventional materials usually used for structural elements in order to achieve lightweight, easily assembled structures.

A typical example of this concept is shown in Fig. 1.1, the Eyecatcher building in Basel, Switzerland. A 15-m-tall, five-story, mobile, lightweight building, it is the tallest multistory GFRP building in the world. Eyecatcher made its first appearance at a building fair, but was then disassembled and reassembled at another location in Basel, Switzerland. The building concept was based on a single-layer load-bearing GFRP envelope integrating structural, building physical and architectural functions. Three GFRP frames composed of adhesively-bonded assembled sections were used as the main load-bearing structure, with translucent aerogel-filled GFRP sandwich walls being placed between the GFRP frames.

The Pontresina bridge (shown in Fig. 1.2) is a temporary lightweight pedestrian bridge, installed each year in the autumn and removed in the spring. Two 2×12.50-m truss girder spans, with adhesively-bonded joints in one span (fully load-bearing) and bolted joints in the other span, were used and additional structural safety was provided by a redundant truss and joint configuration.

Composites were introduced in the aerospace sector long before any other engineering domain since they offer multiple advantages such as high strength and stiffness at light weight—an asset directly connected to cost reduction or higher

Fig. 1.2 The Pontresina
footbridge

transferring capacity—and impressive thermal stability, needed when structures operate in extremely aggressive environments as in space. Numerous applications of "advanced" composite materials in the aerospace industry can be found. Back in 1940 the De Havilland Aircraft Company introduced the Mosquito, an unarmed bomber aircraft made from plywood, mainly Ecuadorian balsawood, sandwiched between Canadian birch to achieve the desired properties. Since then the use of FRP composites in fighter and commercial airplanes has progressively increased. The Airbus 380 is the first commercial airliner with the central wing box made of CFRP, with composites representing a total of 25% of the materials used overall. The main material used was GLARE, a relatively newly developed glass-reinforced fiber metal laminate composed of several very thin layers of aluminum interspersed with layers of glass fibers, bonded together with an epoxy matrix.

Although for the above-mentioned examples and numerous other applications the substitution of conventional materials by FRP composites has proved successful, this practice prevents engineers from taking advantage of one of the most attractive characteristics of composite materials. These "advanced" materials allow engineers to adopt a different approach to design problems, propose alternative design concepts (based on the free formability and lightweight characteristics of composites) and redesign structures. It is thanks to this free formability concept and the superior specific mechanical properties offered by FRP composites that the wind industry grew so rapidly during the last quarter of the twentieth century and is still growing by using hybrid tailor-made materials in order to meet the requirements for today's "huge" wind turbine rotor blades, see Fig. 1.3.

Multifunctional structural elements can also be constructed using composite materials. According to [3] multifunctional composite materials (MFCMs) are defined as the structural composite materials that are designed to perform more than one other non-structural function. Some of the multiple subsystem functions include thermal management, damping, electrical energy generation and storage, sensing radiation shielding and health monitoring.

The integration of solar cells in composite sandwich panels for roof applications [4] to produce load-bearing, energy-producing lightweight structural

Fig. 1.3 7.5 MW Enercon E126 wind turbine. Hub height 135m, rotor diameter 126m. (Photo by jfz, licensed under a Creative Commons Attribution 3.0)

elements, Fig. 1.4, or the use of PU foam in a composite roofing system for noise and heat insulation together with the stabilizing effect of the foam in the sandwich [5], see Fig. 1.5, are typical examples of multifunctional sandwich structures.

The sandwich roof construction in Fig. 1.5 integrates static, building physical and architectural functions, allowing the prefabrication of the entire roof in only four lightweight elements that were easily transported to the site and rapidly installed, see Fig. 1.6. The cutting of foam blocks with a computerized numerical control machine and adhesive bonding proved to be advantageous procedures for the fabrication of the complex roof shape, without the use of expensive molds.

The degree of prefabrication guaranteed a high-quality fabricated composite structural element, quality control of fabrication since it can be performed in well-controlled laboratory conditions, and results in reduced construction times compared to conventional building procedures. In special cases, bridge structures for example, the advantage is even greater since prefabrication and fast installation also mean reduced traffic interference. The lightweight characteristic of composite materials can in this case also offer considerable savings in dead loads and therefore allow the bridge widening (on the same foundation) to accommodate more traffic.

Fig. 1.4 Structural sandwich
element with encapsulated
photovoltaic cells

1.3 Why Fatigue Is Important

Although composite materials are designated as being fatigue-insensitive, especially when compared to metallic ones, they also suffer from fatigue loads. The use
of composite materials in a wide range of applications obliged researchers to
consider fatigue when investigating a composite material and engineers to realize
that fatigue is an important parameter that must be considered in calculations
during design processes, even for structures where fatigue was not traditionally
considered an issue. Although composites were initially used as replacements for
"conventional" materials such as steel, aluminum or wood, and later as
"advanced" materials allowing engineers to adopt a different approach to design
problems, the fatigue behavior of composite materials is different from that of
metallic materials. Therefore, the already developed and validated methods for the
fatigue life modeling and prediction of "conventional" materials cannot be
directly applied to composite materials. Moreover, the large number of different
material configurations resulting from the multitude of fibers, matrices, manu-
facturing methods, lamination stacking sequences, etc. makes the development of

Fig. 1.5 Novartis Campus Main Gate Building with GFRP sandwich roof, view from the south

Fig. 1.6 Roof assembly at construction site

a commonly accepted method to cover all these variances difficult. As stated in [6] "obviously, it is difficult to get a general approach of the fatigue behavior of composite materials including polymer matrix, metal matrix, ceramic matrix composites, elastomeric composites, GLARE, short fiber reinforced polymers and nano–composites".

There is a long list of reasons why fatigue is critical for composite structural components or structures:

- Composites are used for critical structural components and nowadays they participate as a material candidate equal to the traditionally used steel, aluminum or concrete, in emerging structures that must bear significant fatigue loads during operation, such as airplanes, wind turbine rotor blades, leisure boats, foot and vehicular bridges etc. This development changes the common perception concerning the sensitivity of each structure to fatigue. For example, whereas a concrete road bridge is normally not fatigue-sensitive since the dead loads are significantly higher than the live loads, fatigue becomes an issue for a light-weight composite bridge.

• Unidirectional composite materials are generally brittle and behave linearly under load. Since their failure is sudden, without any prior notice, An understanding of their fatigue behavior and prediction of their fatigue life are of major importance.

• An understanding of composite material fatigue behavior is also valuable for the improvement of product development practices. The hitherto followed product development practice was based on an iterative process whereby a prototype was built and tested against real, or realistic, loading patterns. However, this process is costly and time-consuming. The ability to simulate the fatigue behavior of the material, structural component and/or structure reduces the cost and allows the development of a wider range of products without the need for increasing the number of physical prototypes.

• The durability of composite structures is also an important factor. The danger of evaluating durability on the basis of static strength calculations is that the durability impact of cyclic loads is likely to be disregarded. The introduction of fatigue life prediction methodologies into durability simulation procedures allows the assessment of durability performance early in the product development process and the establishment of clear recommendations for guiding major design choices.

1.4 Introduction to the Fatigue of FRP Composites

The scientific community long ago identified fatigue as a critical loading pattern. Back in 1829 the German mining engineer W.A.S. Albert was the first to carry out fatigue tests on metallic conveyor chains [7] and later reported his observations. Subsequently, numerous failures that could not be explained on the basis of known theory were attributed to fatigue loading. With the development of the railways in the mid-nineteenth century, the failure of wagon axles was such a frequent occurrence that it attracted the attention of engineers. Between 1852 and 1870 another German engineer, August Wöhler, realized the first extended experimental program on the fatigue of metallic materials [7]. The program comprised full-scale fatigue tests on wagon axles but also specimen tests under cyclic loading patterns of tensile, bending and torsional loads. Wöhler constructed a test rig on which he could test wagon axles under bending moments developed by loads suspended from the ends of the axles. The developed stresses were recorded together with the number of rotations up to failure. The results were drawn on the "number of cycles (N) vs. stress (σ)" plane to formulate the first S–N curve, which, however, was restricted to the representation of experimental data, without proposing any mathematical formulation to describe this behavior.

These first attempts to analyze the fatigue behavior of materials and structures were based on experience with constructions operating under real loading conditions. Failures that could not be explained by existing theories were designated fatigue failures. As from 1850, engineers recognized fatigue as a critical loading pattern that could be the reason for a significant percentage of structural failures and it was thereafter widely accepted that fatigue should not be neglected. However, as mentioned in the work of Schütz [8], knowledge concerning certain methods was very advanced in one location, while a few kilometers away it was nonexistent. It was not until 1946, when the term fatigue was incorporated in the dictionary of the American Society for Testing and Materials (ASTM), that the E9 committee was founded to promote the development of fatigue test methods [9].

During the following years, numerous experimental programs were conducted for the characterization of the fatigue behavior of several structural FRP composite materials of that time. As technology developed and new test frames and measuring devices were invented, it became more and more straightforward to conduct complex fatigue experiments and measure properties and characteristics, something which some years earlier would not have been possible. As a result, almost all failure modes of FRP composites were identified and many theoretical models were established for modeling and eventually predicting the fatigue life of several different material systems.

1.4.1 Fatigue Damage

The mechanical behavior of composite materials is influenced by a variety of parameters. The homogeneity and quality of the manufactured materials are essential for their structural integrity. Inherent defects like wrinkles, fiber misalignments, and voids that can easily be introduced during the fabrication process can constitute potential damage initiation points and rapidly develop to failure mechanisms like matrix cracking, fiber breakage, debonding, transverse-ply cracking, interface cracking, etc. These failure mechanisms occur sometimes independently and sometimes interactively and are sources of microbuckling, translaminar crack growth and delaminations [10] that can yield catastrophic failure.

The degree of damage in a polymer matrix composite material can be followed by measuring the decrease of a relevant damage metric, usually the residual strength or residual stiffness [11]. A theory based on the residual strength degradation assumes that damage is accumulated in the composite and failure occurs when the residual strength decreases to the maximum applied cyclic stress level, see Fig. 1.7. The fatigue theories based on strength degradation present three major weaknesses:

- The remaining fatigue life cannot be assessed by non-destructive evaluation since the theories are based on a damage metric that necessitates the failure of the material in order to derive it.
- Residual strength degradation is not a sensitive measure of damage accumulation as it changes very slowly until close to failure when it decreases rapidly.

Fig. 1.7 Degradation of
composite strength and
stiffness during constant
amplitude fatigue loading

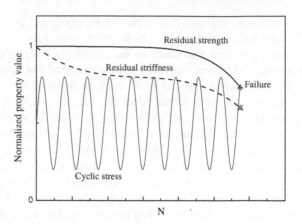

This behavior led Chou and Croman [12] to introduce the term "sudden-death phenomenon" to describe composite fatigue.

- Extensive experimental characterization is needed for each laminate and material system in order to establish a comprehensive database for residual strength.

Attempts to overcome the weaknesses of strength degradation theories have involved damage metrics that can be measured using non-destructive techniques, such as stiffness degradation. Moreover, stiffness changes are usually greater than residual strength changes during composite fatigue life, as schematically shown in Fig. 1.7.

1.4.2 Mechanical Properties: Experimental Investigation of FRP Composites

An intimate knowledge of the stiffness, strength and fatigue performance of the materials used for a structure is crucial for efficient design and these mechanical and elastic material properties can be estimated via the experimental investigation of specimens or structural components. The measured properties undergo statistical analysis and safety factors are applied according to the application and environment.

The typical properties required for the analysis of orthotropic media like FRP composites can be defined by relatively simple tensile, compressive and shear experiments, which are normally performed (when possible) according to well-established standards that define specimen dimensions, pre-treatment, experimental parameters, e.g., loading rate, and fatigue frequency and the desired contents of an experimental report. Typical tensile unidirectional CFRP specimens with fibers along the loading direction (the narrow specimens) and transverse fibers (the wide ones) are presented in Fig. 1.8 with sandwich panels from the same material employed as skin and honeycomb core, used for bending experiments.

As mentioned in [2], the motivation for an experimental investigation is directly related to the aim of the experimental output. Small scale (specimens), components (structural elements) or full-scale (entire structures) experiments can be performed depending on the desired behavior characterization.

According to [2] specimen investigation for research purposes usually involves experiments on standardized specimens to examine material behavior and characterization of the damage development process. Representative examples can be found in [13–15], where the authors attempted to characterize laminate behavior and explain the macroscopically measured strength or stiffness degradation based on observation of failure surfaces, or in [16–18], where fractography was used to describe the Mode I interlaminar fracture toughness of multidirectional laminates [16], angle-ply carbon/nylon laminates [17], and unidirectional and angle-ply glass/polyester DCB specimens [18].

It can also involve experiments on standardized specimens in order to characterize the material and develop theoretical models for the description of its behavior. Unlike the previous category, this kind of testing program focuses mainly on macroscopic observations and data acquisition, e.g., measurement of stiffness degradation, residual strength or the derivation of S–N curves. Representative examples can be found in [19] and [20], where the authors created their own fatigue databases to develop multiaxial fatigue failure criteria, or in [21], where the authors created their database in order to evaluate existing multiaxial fatigue failure theories.

Another aim of specimen investigation for engineering purposes—including tests on predetermined standardized specimens of different materials and/or different configurations—is material selection and optimization of specimen configuration. Often, experimental programs of this type are carried out within the framework of an industrial application, such as the wind turbine rotor blade industry for which a significant number of fatigue databases exist, e.g., [22, 23], or the aerospace industry.

Furthermore, component investigation for research purposes aims to develop analytical models for the modeling and subsequent prediction of the fatigue life of the examined components. Representative work in this category includes that on adhesively-bonded and bolted joints used as structural components in a wide range of applications, e.g., [24–28].

Specimen/component/full-scale investigation for design verification is also used involving experimental programs performed to validate the design of a structure, normally based on quasi-static load cases. The design verification is usually performed by applying representative constant amplitude fatigue loading, e.g., at the serviceability limit state, defined as the load that produces a maximum deformation equal to the span/600 for an FRP bridge deck in [29], or by using accelerated testing for simulation of the long-term behavior of the examined materials. Full-scale investigation is performed to validate the design of a prototype (verify that its lifetime is at least as long as expected) and measure the damage that develops

Fig. 1.8 Typical carbon/
epoxy (CFRP) for tension
loading and CFRP/
honeycomb sandwich
specimens for bending
experiments

throughout this lifetime. Thanks to full-scale investigation, such factors as size or free edge effects (occurring when shifting from the specimen to the full-scale application) are eliminated and credible results regarding the fatigue life of the final structure can be obtained. The high cost and time limitations are the disadvantages of this type of experiment.

The design of the experimental set-up is critical. The thin laminates usually examined as being representative of lightweight composite structures are susceptible to instabilities. As shown in Fig. 1.9, an adhesively-bonded joint can undergo buckling under compression loading if it is inappropriately designed.

A series of antibuckling fixtures have been designed for the protection of thin composite laminates under compression loads with, as mentioned in [10], contradictory results however.

Estimation of shear properties is also problematic since it is difficult to apply loads in order to develop a pure shear stress field in the examined materials. Although for estimation of laminate shear properties the ±45° tensile test [30] can be used, the shear modulus is most successfully measured [10] using the V-notch beam method standardized in ASTM D5379 [31]. This fixture is shown in Fig. 1.10 with a failed V-notched epoxy adhesive specimen.

There are additional parameters that control fatigue behavior and must be adjusted in order to appropriately design fatigue experiments, the most important being the loading pattern, loading frequency, control mode, stress ratio, waveform type, temperature and humidity of the testing environment etc. A complete list of the effects of each of these parameters is given in [2]. Only limited information exists in the form of standards concerning the fatigue investigation of FRP composite materials. ASTM D3479 [32] refers to the tension-tension fatigue of polymer matrix composite materials but gives only vague guidelines without prescribing clamping procedures, loading frequency and methods for data reduction. ISO 13003 [33]for the tension-tension fatigue defines general procedures for fatigue investigation involving all modes of testing machines, but again offers only limited suggestions regarding data evaluation procedures.

Fig. 1.9 Adhesively-bonded
GFRP joint buckled under
compression

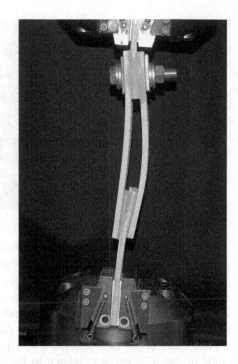

1.4.3 Fatigue Data Manipulation

The adoption of methods for the interpretation of static and fatigue data in order to
interpolate between experimental data (modeling) and extrapolate beyond that for
the prediction of the expected material behavior is a demanding task that depends
on the examined material and thermomechanical loading conditions. Deterministic
or stochastic theoretical models can be employed for this purpose. The use of the
selected models (S–N curves, constant life diagrams, residual strength models,
residual stiffness models, etc.) permits interpretation of the fatigue data and estimation of the fatigue life of the materials, theoretically under any applied loading
pattern.

August Wöhler, as far back as the 1850s, conceived the idea of representing
cyclic stress against the number of cycles to failure in order to quantify the results
of his experimental program. The only input required with regard to experimental
data consists of pairs of numbers of cycles up to failure and the corresponding
alternating stress or strain parameter. The S–N or ε-N curve of the material is then
determined under the applied loading condition. However, another half century
passed before the introduction of the first mathematical model to describe this
relationship. In 1910 Basquin stated that material increases as a power law when
the external load amplitude decreases. The Basquin relationship can take the forms
of the Eq. 1.1 or 1.2:

$$N\sigma^m = Const \tag{1.1}$$

Fig. 1.10 Fixture for V-notched specimen shear investigation

$$\sigma = \sigma_o N^{-1/k} \tag{1.2}$$

where S can be any stress variable such as cyclic stress amplitude, σ_a, maximum cyclic stress, σ_{max}, or cyclic stress range, $\Delta\sigma$ and N is the number of cycles the material can sustain until failure under the corresponding stress value. C, S_o, m and k are model parameters that can be easily estimated by linear regression of the above relationships to the experimental data.

The value of the exponent (m or k, depending on used model) that denotes the slope of the S–N curve is related to the examined material [2]. It is now documented that for fiber-reinforced polymer composite materials this exponent (k) ranges between 7 and 25, being higher (less steep curve) for unidirectional carbon and lower (steeper curve) for multidirectional glass fiber composites. Although laws of this kind are empirical and have no direct physical meaning, they follow the accumulation of the microscopic damage to the examined material that they finally describe [34]. A value of 10 was obtained for a wide and disparate range of GFRP materials, including molded reinforced thermoplastics [35]. As mentioned in [35] the actual life of any given sample of composite type is the manner in which other mechanisms such as transverse-ply cracking in 0/90 laminates, or local resin cracking in woven cloth composites, modify the damage accumulation rate in the load-bearing fibers. Nevertheless, a single mechanistic model that includes randomly reinforced dough-molding compounds and injection-molded thermoplastics as well as woven and non-woven laminates is unlikely to be developed.

The method of estimating S–N curves based on constant amplitude fatigue data is schematically shown in Fig. 1.11. Three different tests are presented, each corresponding to different stress levels and resulting in different numbers of cycles to failure. As expected, the lower the stress level, the longer the fatigue life of the examined material. Interpolation between the collected fatigue data results in the S–N curve of the material under the selected fatigue conditions—R–ratio, frequency, environment etc. The debate concerning the representation of the constant amplitude fatigue data that began a century ago, with Basquin's equation, is still alive and well today. Yet in 2010 there is no universal theoretical model able to

Fig. 1.11 Schematic
representation of S–N curve
derivation

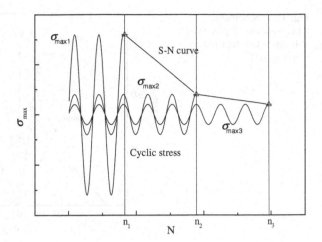

accurately describe the constant amplitude fatigue behavior of any composite material (in terms of S–N curve) under any thermomechanical loading condition. It is accepted that a Log–Log representation (corresponding to the Basquin equation) is sufficiently accurate for most commonly used composite laminates. However, other methods have been proposed to also take into account the low cycle fatigue range, e.g., [36], and consider the statistics of the data sample, [36–38].

The use of quasi-static strength data for the derivation of fatigue curves (as fatigue data for 1 or ¼ cycle) is also arguable. No complete study on this subject exists. Previous publications, e.g., [39], showed that quasi-static data should not be a part of the S–N curve, especially when they have been acquired under strain rates much lower that those used in fatigue loading. The use of quasi-static data in the regression leads to incorrect slopes of the S–N curves as shown in [39]. On the other hand, excluding quasi-static data, although it improved the description of the fatigue data, introduced errors in lifetime predictions when the low cycle regime is important, as for example in the case of loading spectra with a few high-load cycles.

Data from several S–N curves obtained under different loading patterns (tensile, compressive and combinations of both) can be plotted together on the stress mean-stress amplitude (σ_m–σ_a)-plane and form the so-called constant life diagram (CLD) in order to take into account the effect of mean stress on material fatigue life. This diagram helps to derive S–N curves under different R-ratios from those derived experimentally. Constant life (CL) lines connecting data from different S–N curves are then estimated. The CL lines can be linear or non-linear depending on the selected CLD formulation. Unlike constant life diagrams for metals, for which compression stresses were significant since they acted only to close the failure-driving fatigue crack, both left and right quarter cycles are necessary and important for composite materials. Several different models have been introduced. The most recent and commonly used CLDs are in [40]. Starting with the basic idea of the symmetric and linear Goodman diagram and non-linear Gerber equation, different modifications were proposed to cover the peculiarities of composite material

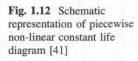

Fig. 1.12 Schematic representation of piecewise non-linear constant life diagram [41]

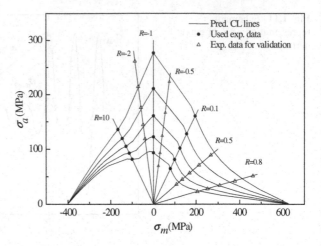

behavior. A typical piecewise non-linear [41] constant life diagram is shown in Fig. 1.12.

1.5 Phenomenological Fatigue Failure Theories

Mathematical models have been developed to describe fatigue damage analytically and eventually predict the fatigue lifetime of FRP composite materials. The ideal fatigue theory is described by Sendeckyj in [11] as one based on a damage metric that accurately models the experimentally observed damage accumulation process, considers all pertinent material, test and environmental variables, correlates the data for a large class of materials, permits the accurate prediction of laminate fatigue behavior from lamina fatigue data, is readily extendable to two-stage and spectrum fatigue loading and takes data scatter into account. These requirements cannot be met simultaneously for many reasons [11] and theoretical models that address only some of them have been introduced. For predicting the fatigue life of structural components made of composites, at least two alternative design concepts could be used: the *damage-tolerant* (or *fail-safe*) and the *safe-life* design concepts. In the former it is assumed that a damage metric, such as crack length, delamination area, residual strength or stiffness, can be correlated to fatigue life via a valid criterion. The presence of damage is permitted as long as it is not critical— i.e., it cannot lead to sudden failure. In the latter—*safe-life* design situations— cyclic stress or strain is directly associated to operational life via the S–N or ε-N curves. The structure is allowed to operate since no damage is observed, e.g., before the initiation of any measurable cracks. Although this design approach ensures the use of safe structures, it considerably increases their cost since it requires very low design values, below the estimated fatigue threshold observed in fracture mechanics experiments on FRP materials.

One of the broadest groups of theoretical models, representing *damage-tolerant* design concepts, comprises the "phenomenological fatigue failure theories", also referred to as "empirical fatigue theories". Models of this type are based on the definition of reliable S–N curves and constant life diagram formulations that are used to estimate allowable numbers of cycles to failure under any given loading pattern from constant to variable amplitude. For most practical cases however, designers require models of behavior that can predict failure under realistic load combinations that yield realistic combinations of stresses, rather than under the uniaxial stress states that usually develop during laboratory experiments. Multiaxial fatigue failure criteria have been developed to take multiaxial fatigue into account [19, 20, 42–47]. Most of the aforementioned examples in the literature concentrate mainly on the introduction and validation of fatigue failure criteria suitable for constant amplitude multiaxial proportional stress fields without addressing the problem of life prediction under irregular load spectra.

An exception to the above is presented in a series of publications [48–51] in which a complete life prediction methodology, for even a 3D stress state, is established. Their approach is based on a "ply-to-laminate characterization" scheme, dealing with damage accumulation under variable amplitude loading by adopting appropriate residual strength engineering models. Another paradigm of a complete life prediction methodology under irregular plane stress histories and an experimental database for a glass fiber-reinforced polyester (GFRP) [0/45]$_S$ laminate are presented in Chaps. 2, 6 and 7.

1.6 Conclusions: Future Trends

Many research efforts have focused on investigation of the fatigue behavior of FRP composite materials. Almost all of the encountered failure modes have been recognized and are well documented in the literature and their effect on the fatigue life of the examined material has been quantified. However several problems regarding the fatigue behavior and life modeling/prediction of FRP composite materials remain unresolved. Some of these are listed below:

The fatigue behavior of composites becomes more complicated when the influence of temperature is also included. To the authors' knowledge, there is no method in the literature that can address thermomechanical loads and predict the fatigue behavior for general thermomechanical loading patterns, although much research effort has been devoted to the characterization of the fatigue behavior of adhesively-bonded composite joints and composite laminates in different temperature and humidity environments. A small number of modeling approaches have been published, e.g., [52, 53]. However, to accommodate a significant number of parameters that affect the fatigue life of FRP joints or laminates, these phenomenological models adopt a great many assumptions and their applicability could therefore not be validated for the data concerning different material systems.

The mean stress effect on the fatigue life of composite materials and structures is very pronounced. As explained in a number of publications [54–56], and Chap. 4 of this volume, this effect can be accurately modeled using the so-called constant life diagrams. The most recent CLD formulations were presented in 2007 by Kawai and Koizumi [55] and 2010 by Vassilopoulos et al. [41]. Although from a theoretical point of view the classic representation of the CLD is rational, it presents a deficiency when seen from the engineering point of view. This deficiency is related to the region close to the horizontal axis, which represents loading under very low stress amplitude and high mean values with a culmination for zero stress amplitude ($R = 1$). The classic CLD formulations require that the constant life lines converge to the ultimate tensile stress (UTS) and the ultimate compressive stress (UCS), regardless of the number of loading cycles. However, this is an arbitrary simplification originating from the lack of information about the fatigue behavior of the material when no amplitude is applied. In fact, it is preferable that this type of loading should not be considered as fatigue loading, but rather as material creep (constant static load over a short or long period). Although modifications have been introduced to take into account the time-dependent material strength, their integration into CLD formulations requires the adoption of additional assumptions, see e.g., [54, 57].

Engineering structures like bridges or wind turbine rotor blades can undergo between 10^8 and 10^9 loading cycles during their operational lifetime. On the other hand, the usual range of recorded fatigue data lies between 10^3 and 10^6–10^7 loading cycles. It is obvious that a model to extrapolate material behavior is needed in order to estimate the fatigue life of materials in real loading environments. However, existing fatigue models usually fail to accurately model this effect since they have no information concerning this cycle regime. Accelerated testing is used in laboratories to compare the predicting ability of existing fatigue models to realistic loading, but this type of testing masks the effect on fatigue life of the existence of low cycles in real operational loads. Recently a number of researchers tried to perform high-cycle fatigue tests [22, 58, 59]. Mandel's group [58] performed tests on very small-diameter impregnated strands with only sufficient fibers to be representative of the behavior of larger specimens. The load was applied using low-frequency audio speakers (woofers) as actuators that can handle frequencies in the range of 300 Hz. Hosoi et al. [59] tested standardized quasi-isotropic carbon fiber-reinforced plastic laminates with a stacking sequence of $[45/0/-45/90]_S$ and reaching a frequency of 100 Hz over more than 10^8 loading cycles. According to the authors, frequency does not affect fatigue behavior as long as the temperature remains well below the glass transition temperature of the examined material. These two publications prove that it is possible to perform high frequency tests and validate existing models (or develop new ones) for interpretation of fatigue data in the high-cycle fatigue regime as well.

It is also a known fact that engineering structures used in open-air applications undergo more complicated loading spectra than the constant or block loading conditions usually applied in laboratories and thereafter used for the description of the fatigue behavior of the examined material. The behavior of composite materials is also known to be affected by load sequence [60], overloads [61, 62] and changes in loading. A considerable amount of time has been spent on addressing this problem and quantifying the effects of spectrum loading on the fatigue behavior of several material systems, e.g., [22, 23, 63], and structural components [61, 62]. Other researchers worked on the development of accelerated testing methods in order to simulate lengthy loading spectra, e.g., [64, 65].

Innovative measuring techniques can also assist in the health monitoring of a structure during operational life in order to achieve fail-safe, cost-effective designs. Acoustic emission (AE) has been used in the past in combination with artificial neural networks to characterize damage in CFRP composite laminates, e.g., [66], or for the assessment of the normal and shear strength degradation in FRP composite materials during constant and variable amplitude fatigue loading, e.g., [67]. Optical Fiber Bragg Grating (FBG) sensors have been used for the characterization of the residual stresses in single-fiber composites [68] and the monitoring of hygrothermal aging effects in epoxy resins [69].

As mentioned, a drawback of stiffness and strength degradation theories that focus on material properties in the principal loading direction is their inability to take into account multiaxial property degradation. However, efforts have recently been devoted to the development of models that also take other material elastic constants into account, such as degradation of the Poisson's ratio, transverse stiffness and in-plane shear modulus [70] or to the use of strength degradation theories together with phenomenological modeling for estimation of material properties during fatigue loading [71].

Within the next few years the scientific community is expected to come up with a credible methodology for the life prediction of engineering FRP structures which will allow rapid and reliable prototyping without the need for the fabrication and testing of numerous physical prototypes. Therefore the commercialization of a wide range of reliable products will be feasible within shorter time periods and at reduced costs. This process is well assisted by developments in computer science, where computational power has been increased dramatically during the last decade. Although this increase is now less rapid, today it is feasible to use even personal computers to "run" progressive damage models incorporated in finite element analysis software and estimate the fatigue life of an entire structure within hours.

Some of the aforementioned topics are addressed in the following chapters of this volume. Research on others can be traced in the interesting papers that are published every year. With more than 500 scientific publications in international journals per year during the last decade, the fatigue of composites seems to be a very hot topic in engineering and materials science.

References

1. S.A. Paipetis, *Science and Technology in Homeric Epics* (Springer, Berlin, 2008), pp. 181–203
2. A.P. Vassilopoulos, Introduction to the fatigue life prediction of composite materials and structures: Past, present and future prospects, in *Fatigue Life Prediction of Composites and*

Composite Structures, ed. by A.P. Vassilopoulos (Woodhead Publishing Limited, Cambridge, 2010), pp. 1–44
3. S.P. Rawal, Multifunctional composite materials and structures. Comp. Compos. Mater. **6.06**, 67–86 (2003)
4. T. Keller, A.P. Vassilopoulos, B.D. Manshadi, Thermomechanical behavior of multifunctional GFRP sandwich structures with encapsulated photovoltaic cells. J. Compos. Constr. **14**(4), 470–478 (2010)
5. T. Keller, C.H. Haas, T. Vallee, Structural concept, design and experimental verification of a GFRP sandwich roof structure. J. Compos. Constr. **12**(4), 454–468 (2008)
6. C. Bathias, An engineering point of view about fatigue of polymer matrix composite materials. Int. J. Fatigue **28**(10 SPEC. ISS.), 1094–1099 (2006)
7. J.A. Collins, *Failure of Materials in Mechanical Design* (Wiley, New York, 1993)
8. W. Schütz, A history of fatigue. Eng. Fract. Mech. **54**(2), 263–300 (1996)
9. J.T. Fong, What is fatigue damage?, in *Damage in Composite Materials, ASTM STP 775*, ed. by K.L. Reifsnider (American Society for Testing and Materials, West Conshohocken, PA, 1982), pp. 243–266
10. P. Brondsted, H. Lilholt, A. Lystrup, Composite materials for wind power turbine blades. Annu. Rev. Mater. Res. **35**, 505–538 (2005)
11. G.P. Sendeckyj, Life prediction for resin-matrix composite materials, in *Fatigue of Composite Materials*, ed. by K.L. Reifsnider (Elsevier, Amsterdam, 1991), pp. 431–483
12. P.C. Chou, R. Croman, Degradation and sudden-death models of fatigue of graphite/epoxy composites. in Proceedings of the 5th conference on composite materials: Testing and design, ASTM STP 674, (1979), pp. 431–454
13. M.M. Ratwani, H.P. Kan, Effect of stacking sequence on damage propagation and failure modes, in *Composite Laminates, in Damage in Composite Materials, ASTM STP 775*, ed. by K.L. Reifsnider (American Society for Testing and Materials, West Conshohocken, PA, 1982), pp. 40–62
14. H.T. Hahn, Fatigue behavior and life prediction of composite materials. in Composite materials: Testing and design (5th conference). ASTM STP 674. ed. by S.W. Tsai (1979) pp. 383–417
15. T.P. Philippidis, A.P. Vassilopoulos, Stiffness reduction of composite laminates under combined cyclic stresses. Adv. Compos. Lett. **10**(3), 113–124 (2001)
16. H. Chai, The characterization of Mode I delamination failure in non–woven, multidirectional laminates. Composites **15**(4), 277–290 (1984)
17. K. Tohgo, Y. Hirako, H. Ishii, Mode I interlaminar fracture toughness and fracture mechanism of angle–ply carbon/nylon laminates. J. Compos. Mater. **30**(6), 650–661 (1996)
18. F. Ozdil, L.A. Carlsson, Beam analysis of angle–ply laminate DCB specimens. Compos. Sci. Technol. **59**(2), 305–315 (1999)
19. Z. Hashin, A. Rotem, A fatigue failure criterion for fibre–reinforced materials. J. Compos. Mater. **7**(4), 448–464 (1973)
20. T.P. Philippidis, A.P. Vassilopoulos, Complex stress state effect on fatigue life of GRP laminates. Part I, experimental. Int. J. Fatigue **24**(8), 813–823 (2002)
21. M.J. Owen, J.R. Griffiths, Evaluation of biaxial stress failure surfaces for a glass fabric reinforced polyester resin under static and fatigue loading. J. Mater. Sci. **13**(7), 1521–1537 (1978)
22. J.F. Mandell, D.D Samborsky, DOE/MSU composite material fatigue database: Test methods material and analysis, Sandia National Laboratories/Montana State University, SAND97–3002, (online via www.sandia.gov/wind, last update, v. 15.0, 2nd March 2006)
23. R.P.L. Nijssen, OptiDAT–fatigue of wind turbine materials database (2006) http://www.kc-wmc.nl/optimat_blades/index.htm
24. H. Hadavinia, A.J. Kinloch, M.S.G. Little, A.C. Taylor, The prediction of crack growth in bonded joints under cyclic–fatigue loading I. Exp. Stud. Int. J. Adhes. Adhes. **23**(6), 449–461 (2003)

25. Y. Zhang, A.P. Vassilopoulos, T. Keller, Stiffness degradation and life prediction of adhesively–bonded joints for fiber-reinforced polymer composites. Int. J. Fatigue 30(10–11), 1813–1820 (2008)
26. Y. Zhang, A.P. Vassilopoulos, T. Keller, Environmental effects on fatigue behavior of adhesively–bonded pultruded structural joints. Compos. Sci. Technol. 69(7–8), 1022–1028 (2009)
27. J. Schön, T. Nyman, Spectrum fatigue of composite bolted joints. Int. J. Fatigue 24(2–4), 273–279 (2002)
28. R. Starikov, J. Schön, Local fatigue behavior of CFRP bolted joints. Compos. Sci. Technol. 62(2), 243–253 (2002)
29. T. Keller, H. Gürtler, Quasi–static and fatigue performance of a cellular FRP bridge deck adhesively bonded to steel girders. Compos. Struct. 70(4), 484–496 (2005)
30. D3518/D3518 M-94(2007) Standard test method for in-plane shear response of polymer matrix composite materials by tensile test of a ± 45° laminate
31. ASTM D 5379/D 5379 M-05 Standard test method for shear properties of composite materials by the V-notched beam method
32. D3479/D3479 M-96. Standard test method for tension-tension fatigue of polymer matrix composite materials (2007)
33. ISO 13003 Fibre-reinforced plastics – determination of fatigue properties under cyclic loading conditions (2003)
34. F. Kun, H.A. Carmona, J.S. Andrade, H.J. Jr Herrmann, Universality behind basquin's law of fatigue. Phys. Rev. Lett. 100(9), 094301 (2008)
35. B. Harris, A historical review of the fatigue behavior of fiber-reinforced plastics, in *Fatigue in Composites-Science and Technology of the Fatigue Response of Fiber-Reinforced Plastics*, ed. by B. Harris (Woodhead Publishing Limited, Cambridge, 2003), pp. 1–35
36. G.P. Sendeckyj, Fitting models to composite materials fatigue data, in *Test Methods and Design Allowables for Fibrous Composites. ASTM STP 734*, ed. by C.C. Chamis (American Society for Testing and Materials, West Conshohocken, PA, 1981), pp. 245–260
37. J.M. Whitney, Fatigue characterization of composite materials, in Fatigue of fibrous composite materials, ASTM STP 723, (1981), pp. 133–151
38. S. Shimizu, K. Tosha, K. Tsuchiya, New data analysis of probabilistic stress–life (P–S–N) curve and its application for structural materials. Int. J. Fatigue 32(3), 565–575 (2010)
39. R.P.L. Nijssen, O. Krause, T.P. Philippidis, Benchmark of lifetime prediction methodologies. Optimat Blades Technical Report. OB_TG1_R012 Rev.001 (2004) http://www.wmc.eu/public_docs/10218_001.pdf
40. A.P. Vassilopoulos, B.D. Manshadi, T. Keller, Influence of the constant life diagram formulation on the fatigue life prediction of composite materials. Int. J. Fatigue 32(4), 659–669 (2010)
41. A.P. Vassilopoulos, B.D. Manshadi, T. Keller, Piecewise non-linear constant life diagram formulation for FRP composite materials. Int. J. Fatigue 32(10), 1731–1738 (2010)
42. Z. Fawaz, F. Ellyin, Fatigue failure model for fibre–reinforced materials under general loading conditions. J. Compos. Mater 28(15), 1432–1451 (1994)
43. T.P. Philippidis, A.P. Vassilopoulos, Complex stress state effect on fatigue life of GRP laminates. Part II, theoretical formulation. Int. J. Fatigue 24(8), 825–830 (2002)
44. M.J. Owen, J.R. Griffiths, Evaluation of biaxial stress failure surfaces for a glass fabric reinforced polyester resin under static and fatigue loading. J. Mater. Sci. 13(7), 1521–1537 (1978)
45. T. Fujii, F. Lin, Fatigue behavior of a plain–woven glass fabric laminate under tension/torsion biaxial loading. J. Compos. Mater. 29(5), 573–590 (1995)
46. M–.H.R. Jen, C.H. Lee, Strength and life in thermoplastic composite laminates under static and fatigue loads Part I: Experimental. Int. J. Fatigue 20(9), 605–615 (1998)
47. E.W. Smith, K.J. Pascoe, Biaxial fatigue of a glass–fiber reinforced composite. Part 2: Failure criteria for fatigue and fracture, in *Biaxial and Multiaxial Fatigue, EGF3*, ed. by M.W. Brown, K.J. Miller (Mechanical Engineering Publications, London, 1989), pp. 397–421

48. L.B. Lessard, M.M. Shokrieh, Two–dimensional modeling of composite pinned–joint failure. J. Compos. Mater. **29**(5), 671–697 (1995)
49. M.M. Shokrieh, L.B Lessard, C. Poon, Three–dimensional progressive failure analysis of pin/bolt loaded composite laminates, in Bolted/bonded joints in polymeric composites, AGARD CP 590, (1997), pp. 7.1–7.10
50. M.M. Shokrieh, L.B. Lessard, Multiaxial fatigue behavior of unidirectional plies based on uniaxial fatigue experiments–I. Model. Int. J. Fatigue **19**(3), 201–207 (1997)
51. M.M. Shokrieh, L.B. Lessard, Multiaxial fatigue behavior of unidirectional plies based on uniaxial fatigue experiments–II. Exp. Eval. Int. J. Fatigue **19**(3), 209–217 (1997)
52. M. Kawai, N. Maki, Fatigue strength of cross–ply CFRP laminates at room and high temperatures and its phenomenological modeling. Int. J. Fatigue **28**(10), 1297–1306 (2006)
53. M. Kawai, T. Taniguchi, Off–axis fatigue behavior of plain woven carbon/epoxy composites at room and high temperatures and its phenomenological modeling. Compos. Part A–Appl. S **37**(2), 243–256 (2006)
54. H.J. Sutherland, J.F. Mandell, Optimized constant life diagram for the analysis of fiberglass composites used in wind turbine blades. J. Sol. Energy Eng. Trans. ASME **127**(4), 563–569 (2005)
55. M. Kawai, M. Koizumi, Nonlinear constant fatigue life diagrams for carbon/epoxy laminates at room temperature. Compos. Part A–Appl. S **38**(11), 2342–2353 (2007)
56. B. Harris, A parametric constant–life model for prediction of the fatigue lives of fibre–reinforced plastics, in *Fatigue in Composites*, ed. by B. Harris (Woodhead Publishing Limited, Cambridge, 2003), pp. 546–568
57. J. Awerbuch, H.T. Hahn, Off–axis fatigue of graphite/epoxy composites. In: Fatigue of fibrous composite materials. ASTM STP 723, (1981), pp. 243–273
58. J.F. Mandell, D.D. Samborsky, L. Wang, N.K. Wahl, New fatigue data for wind turbine blade materials. J. Sol. Energy Eng. Trans. ASME **125**(4), 506–514 (2003)
59. A. Hosoi, N. Sato, Y. Kusumoto, K. Fujiwara, H. Kawada, High–cycle fatigue characteristics of quasi–isotropic CFRP laminates over 10^8 cycles (initiation and propagation of delamination considering interaction with transverse cracks). Int. J. Fatigue **32**(1), 29–36 (2010)
60. W. Van Paepegem, J. Degrieck, Effects of load sequence and block loading on the fatigue response of fiber-reinforced composites. Mech. Adv. Mater Struct. **9**(1), 19–35 (2002)
61. S. Erpolat, I.A. Ashcroft, A.D. Crocombe, M.M. Abdel–Wahab, A study of adhesively bonded joints subjected to constant and variable amplitude fatigue. Int. J. Fatigue **26**(11), 1189–1196 (2004)
62. S. Erpolat, I.A. Ashcroft, A.D. Crocombe, M.M. Abdel–Wahab, Fatigue crack growth acceleration due to intermittent overstressing in adhesively bonded CFRP joints. Compos. Part A–Appl. S **35**(10), 1175–1183 (2004)
63. T.P. Philippidis, A.P. Vassilopoulos, Life prediction methodology for GFRP laminates under spectrum loading. Compos. Part A–Appl. S **35**(6), 657–666 (2004)
64. Y. Miyano, M. Nakada, J. Ichimura, E. Hayakawa, Accelerated testing for long–term strength of innovative CFRP laminates for marine use. Compos. Part B–Eng. **39**(1), 5–12 (2008)
65. Z. Fawaz, F. Ellyin, Fatigue failure model for fibre–reinforced materials under general loading conditions. J. Compos. Mater **28**(15), 1432–1451 (1994)
66. T.P. Philippidis, V.N. Nikolaidis, A.A. Anastassopoulos, Damage characterization of carbon/carbon laminates using neural network techniques on AE signals. NDT&E Int. **31**(5), 329–340 (1998)
67. T.P. Philippidis, T.T. Assimakopoulou, Strength degradation due to fatigue–induced matrix cracking in FRP composites: An acoustic emission predictive model. Compos. Sci. Technol. **68**(15–16), 3272–3277 (2008)
68. F. Colpo, L. Humbert, J. Botsis, Characterization of residual stresses in a single fibre composite with FBG sensor. Compos. Sci. Technol. **67**(9), 1830–1841 (2007)

69. D. Karalekas, J. Cugnoni, J. Botsis, Monitoring of hygrothermal ageing effects in an epoxy resin using FBG sensor: A methodological study. Compos. Sci. Technol. **69**(3–4), 507–514 (2009)
70. W. Van Paepegem, Fatigue damage modeling of composite materials with phenomenological residual stiffness approach, in *Fatigue Life Prediction of Composites and Composite Structures*, ed. by A.P. Vassilopoulos (Woodhead Publishing Limited, Cambridge, 2010), pp. 102–138
71. T.P. Philippidis, E.N. Eliopoulos, A progressive damage mechanics algorithm for life prediction of composite materials under cyclic complex stress, in *Fatigue Life Prediction of Composites and Composite Structures*, ed. by A.P. Vassilopoulos (Woodhead Publishing Limited, Cambridge, 2010), pp. 390–436

Chapter 2
Experimental Characterization
of Fiber-Reinforced Composite Materials

2.1 Introduction

The anisotropic nature of fiber-reinforced composite materials makes their experimental characterization a complicated task. In general, composites exhibit different properties under tension and compression, under both quasi-static and fatigue loading patterns. The increased number of parameters that affect the material properties, such as loading rate, temperature and humidity, mean stress and load frequency, necessitate the careful and targeted design of an experimental program for the characterization of a composite material in order to use it for a specific application, since it is theoretically impossible to take into account all the parameters and simulate their effect on material behavior. This characteristic of composite materials makes their incorporation in design codes and guidelines difficult.

For the design of a structure based on the static strength or stiffness there are a number of available theories in the literature. It is up to the design engineer to select the failure criterion that fits the requirements of each application and matches the nature of the selected material. However, the criterion selection can be validated only after comparison of the theoretical predictions with some experimental data. It should also be mentioned that the most appropriate criterion for the prediction of material behavior under a given loading pattern can present disadvantages when another material is examined or even for the same material but under different loading conditions.

For fatigue loading, the situation is much more complicated since the material properties change during loading. This difficulty is increased by the fact that this property variation (normally degradation) is not linear since its rate depends on the loading conditions and material status, i.e. how far it is from failure. In general, the inability to accurately model material behavior leads to the adoption of high coefficients of safety, which, in addition to those already used due to the stochastic nature of the fatigue loading leads to the over dimensioning of each structure. It is

A. P. Vassilopoulos and T. Keller, *Fatigue of Fiber-reinforced Composites*,
Engineering Materials and Processes, DOI: 10.1007/978-1-84996-181-3_2,
© Springer-Verlag London Limited 2011

therefore necessary to be able to characterize each material and appropriately model its quasi-static and fatigue behavior in order to develop theories that will assist the design process.

The fatigue behavior of carbon fiber-reinforced plastics (CFRP) has been extensively investigated over the last 40 years thanks to the concentrated effort to develop composite structural components for aeronautical applications. Most aspects of fatigue-related engineering problems, i.e. life prediction, property degradation, joint design etc., were addressed, leading to the adoption of design allowables and the production of a large amount of published data, e.g. [1–5]. However, damage tolerance issues have not been efficiently dealt with [6] for many reasons, the main one being the lack of definition of a generalized damage metric, such as the crack length in metals, that could be used with different lay-ups and material configurations [7]. In addition, the effect of variable amplitude loading on the remaining life and the fatigue behavior of composite materials under complex stress states have only received limited attention.

The structural response to cyclic loads of glass fiber-reinforced plastics (GFRP), extensively used in a number of engineering applications such as leisure boats, transportation, wind turbine rotor blades, helicopter rotor blades and bridge decks, was not investigated to any significant extent until 25 years ago. Due to the amazing growth of the wind energy industry, especially in Europe, much effort has been devoted during the last two decades to establishing fatigue design allowables for GFRP, in particular laminated composites for wind turbine rotor blades. A great deal of experimental data was produced characterizing the fatigue strength of matrix systems such as polyester, epoxies and vinylester reinforced by continuous glass fibers in the form of woven or stitched fabrics and unidirectional roving [8–17]. The effect of both constant and variable amplitude, i.e. spectral loading conditions, was investigated.

However, limited experimental data and design guidelines addressing the complex stress state effect on the fatigue behavior of GFRP laminates have been produced by applying either multiaxial or off-axis loading. Studies [18–22] point out the strong dependency of fatigue response on load direction, as a result of material anisotropy, and indicate the need to continue research in this domain, including the effects of spectral and non-proportional loading.

Typical modern composite structures such as aeronautical vehicles and wind turbine rotor blades are subjected to severe dynamic loads of stochastic and deterministic nature. The stress state in their primary structural elements, in the form of thin and moderately thick shells made of laminated fiber-reinforced polymers, can be assumed plane, i.e. composed of two normal components and an in-plane shear component of the stress tensor. A formidable task for designers is the life prediction of such components subjected to irregular stress histories caused by multiaxial loads of variable amplitude. There are many critical decisions that should be made related to this issue that concern both the experimental charac-terization of relevant mechanical material properties and the establishment of reliable life prediction methods.

In the case of laminated FRP materials, mechanical properties, i.e. elasticity, strength and hygro-thermal property tensors, can be measured for the entire laminate as a homogeneous anisotropic medium, an approach known as "direct characterization". Alternatively, mechanical properties can be measured for each individual layer and then theoretical methods can be used to predict the laminate behavior. There are advantages and disadvantages to both approaches. In the former, results are valid only for the specific lay-up and cannot be used in a laminate optimization design algorithm. In addition, for asymmetric or unbalanced stacking sequences, there are property couplings that cannot be measured appropriately. The latter approach, "ply-to-laminate characterization", although successfully implemented for elastic and hygro-thermal properties, remains an unresolved issue concerning the strength prediction of laminated composites due to difficulties of modeling damage progression and interaction effects. The situation is even more complicated in the case of fatigue strength and life prediction.

Systematic research efforts regarding the fatigue property characterization of GFRP composites have led to the development of magnificent fatigue databases during the last two decades and the establishment of reliable life prediction methodologies [9–16, 20–23]. However, all these experimental investigations focused mainly on the axial property characterization of various laminates, i.e. direct characterization in only one direction of an anisotropic medium. Therefore, this effort and the accumulated experimental results, including constant (CA) and variable amplitude (VA) loading, have limited use when life prediction under irregular, multidimensional stress histories is required.

Experimental results are presented here from a comprehensive program consisting of quasi-static and fatigue tests on straight edge specimens cut at various on- and off-axis directions from a GFRP multidirectional (MD) laminate of $[0/(45)_2/0]_T$ lay-up. The fatigue behavior of off-axis loaded laminates, i.e. complex state of stress in material principal directions, is investigated in depth for several off-axis orientations. This includes the derivation of S–N curves at various R-ratios ($R = \sigma_{min}/\sigma_{max}$), statistical evaluation of fatigue strength results and determination of design allowables at specific reliability levels. Constant life diagrams are extracted for the various off-axis directions and compared with existing data from similar material systems.

2.2 Experimental Program

2.2.1 Structural Fiber-Reinforced Composite Materials: Multidirectional Laminates

The material system was E-glass/polyester, with the E-glass being supplied by Ahlstrom Glassfibre, and the polyester resin, Chempol 80 THIX, by Interchem. This resin is a thixotropic unsaturated polyester and was mixed with 0.4% cobalt

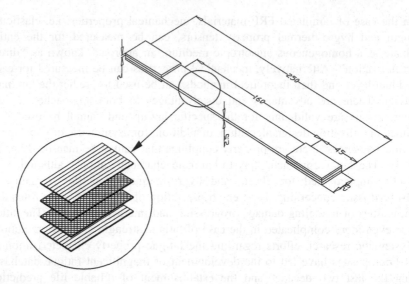

Fig. 2.1 Specimen geometry and stacking sequence. Specimen cut at 90° off-axis is shown

naphthenate solution (6% Co) accelerator and 1.5% methyl ethyl ketone peroxide, MEKP (50% solution) catalyst. A systematic experimental investigation was undertaken, consisting of static and fatigue tests on straight edge specimens cut at various directions from a multidirectional laminate. The stacking sequence of the E-glass/polyester plate consisted of four layers, 2 UD, unidirectional lamina of 100% aligned warp fibers, with a weight of 700 g/m² as outer layers and 2 stitched laminae with fibers along ±45° directions, of 450, 225 g/m² in each off-axis angle.

Rectangular plates were fabricated using a hand lay-up technique and cured at room temperature. Considering the UD layer fibers as being along the 0° direction, the lay-up can be encoded as [0/(45)₂/0]T. Specimens were cut using a diamond wheel at 0°, on-axis, and 15°, 30°, 45°, 60°, 75° and 90° off-axis directions.

The specimens were prepared according to the ASTM 3039/D3039 standard, their edges were trimmed with sandpaper and aluminum tabs were glued at their ends. The specimens were 250 mm long and had a width of 25 mm. Their nominal thickness was 2.6 mm. The length of the tabs, with a thickness of 2 mm, was 45 mm leaving a gauge length of 160 mm for each of the specimens. The specimen dimensions are shown in Fig. 2.1.

Quasi-static and fatigue experiments were performed on the total number of 355 specimens as follows: 31 specimens for static tests to provide baseline data, with both tensile and compressive strengths being obtained, while 277 specimens were tested under uniaxial cyclic stress of constant amplitude for the determination of 17 S–N curves at various off-axis directions and loading conditions and the investigation of the effect of the load interruption on the fatigue life. Additionally, 47 specimens were tested under two different variable amplitude spectra, a modified version of the WISPERX used in the wind turbine rotor blade industry, and a realistic irregular fatigue spectrum derived from aeroelastic calculations.

Table 2.1 Failure stresses of examined material system

	UTS (MPa)		UCS (MPa)	
	Mean value	S.D.	Mean value	S.D.
	Displacement control, 1 mm/min			
0°	244.84 (4)	18.08	216.68 (4)	14.67
30°	130.52 (2)	14.23	145.52 (2)	13.94
45°	139.12 (2)	25.61	106.40 (2)	2.69
60°	117.26 (2)	16.64	99.52 (2)	3.59
90°	84.94 (3)	2.06	83.64 (3)	5.37
Shear	61.38 = (139.12 + 106.40)/2/2			
	Load control, 40 kN/s			
0°	417.49 (5)	74.86		

2.2.2 Quasi-Static Experiments: Off-Axis Strength Prediction

Tensile and compressive experiments were performed for the derivation of the strength of the examined material. The majority of the experiments were performed under displacement-control mode (1 mm/min), while a limited number of tensile experiments on specimens cut at $0°$ were performed under load-control mode with a loading rate of 40 kN/s, much faster than the corresponding loading rate in the displacement-controlled experiments. The results are presented in Table 2.1. The failure stress obtained from the load-controlled experiments was 417.49 MPa, approximately 1.7 times the corresponding strength of 244.84 MPa obtained in the relatively slow, displacement-controlled experiments. This result proves the significant effect of the loading rate on the strength of the examined polymer matrix material.

The symbols X, X', Y and Y' will be used in the following to denote the tensile and compressive strengths along the longitudinal (X) and transverse (Y) material directions. The shear strength is represented by S and is considered to be the same under positive or negative shear stresses, according to the sign convention, a hypothesis that has been experimentally validated in the past and is assumed safe for technical applications and shear strengths along the principal planes of orthotropic media or media with higher symmetry.

The static strengths of the material along different directions are presented in Fig. 2.2. Theoretical predictions provided by the Failure Tensor Polynomial (FTP) quadratic failure criterion [24] are also shown in this figure by a solid line. The form of the failure criterion used is:

$$\frac{\sigma_1^2}{XX'} + \frac{\sigma_2^2}{YY'} - \frac{\sigma_1\sigma_2}{XY} + \sigma_1\left(\frac{1}{X} - \frac{1}{X'}\right) + \sigma_2\left(\frac{1}{Y} - \frac{1}{Y'}\right) + \frac{\sigma_6^2}{S^2} - 1 = 0 \qquad (2.1)$$

where the Tsai-Hahn version of the quadratic failure criterion was used with the interaction term taking the form:

$$F_{12} = -0.5\sqrt{F_{11}F_{12}} \qquad (2.2)$$

Fig. 2.2 Off-axis static strength of examined GFRP laminate

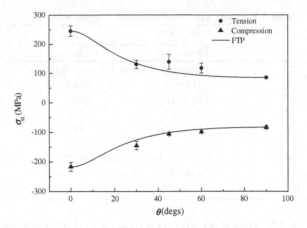

and the shear strength is considered equal to half of the tensile strength of specimens cut at 45° off-axis, as presented in Table 2.1.

With σ_i, $i = 1$, 2, 6 denoting the in-plane stress tensor components in the principal coordinate system (PCS) of the multidirectional laminate and σ_x the applied normal stress at an off-axis angle , the following transformation equation is valid:

$$\sigma_1 = \sigma_x \cos^2 \theta$$
$$\sigma_2 = \sigma_x \sin^2 \theta \qquad (2.3)$$
$$\sigma_6 = \sigma_x \sin \theta \cos \theta$$

The strength of the material along each off-axis angle direction can be calculated by replacing the stresses σ_1, σ_2 and σ_6 in Eq. 2.1 and solving it for σ_x.

The static strength under both tension and compression was the highest for the 0° MD specimens, while is considerably and continuously decreased with higher off-axis angles. The peculiar behavior of the 45° specimens, exhibiting higher static strength than the 30° specimens, was to be expected and was related to the stacking sequence of the laminate under consideration. When specimens are cut at 45°, the fibers of the stitched (±45) layers become on-axis and the prevailing failure mode shifts from matrix-dominated, observed in 30° and 60° off-axis specimens, to partially fiber-controlled. Therefore, these fibers bear most of the load instead of the matrix. This effect was not observed under compressive loads since in that case the matrix is the dominant component and therefore fibers of the ±45 layers do not significantly contribute to the strength of the material. In general, a mixed failure mode was observed for all the off-axis specimens.

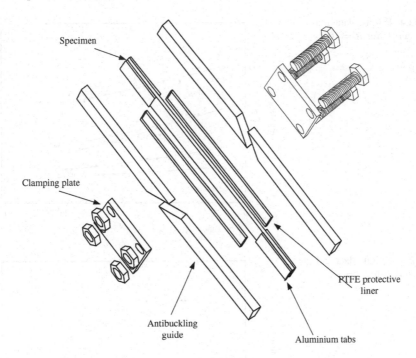

Specimen

Clamping plate

RTFE protective
liner

Antibuckling
guide

Aluminium tabs

Fig. 2.3 Antibuckling device

2.2.3 Fatigue Experiments: Derivation of S–N Curves

Cyclic experiments of sinusoidal waveform and constant amplitude were also carried out in the same machine under load control. In total, 17 S–N curves were determined experimentally along various off-axis loading directions, under four different stress ratios, namely, $R = 10$ representing compression-compression loading (C-C), $R = -1$ representing tension-compression reversed loading (T-C) and $R = 0.1$ and $R = 0.5$ representing tension-tension fatigue loading (T-T). The frequency was kept constant at 10 Hz for all the tests as no appreciable temperature increase was detected during cycling under various loading conditions.

Loading was continued until specimen ultimate failure or 10^6 cycles, whichever occurred first. In particular, for the on-axis specimens, 0°, under reversed loading, $R = -1$, loading was continued for up to 5×10^6 cycles. Specimens that did not fail, and were removed after a specific number of cycles were marked as "runouts". For all the tests with compressive cycles, the antibuckling device shown in Fig. 2.3 was used. At least three specimens were tested at each one of the four or five stress levels. All tests were conducted at room temperature, 18°–22°C.

Uniaxial tests on specimens cut off-axis from principal material directions were performed to induce complex stress states in the principal coordinate system (PCS), according to Eq. 2.3. The biaxiality ratios σ_2/σ_1 and σ_6/σ_1 as a function of take values equal to $\tan^2\theta$ and $\tan\theta$, respectively.

Fig. 2.4 Fatigue data and S–N curves for $R = 10$

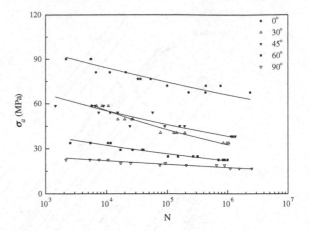

Fig. 2.5 Fatigue data and S–N curves for $R = -1$

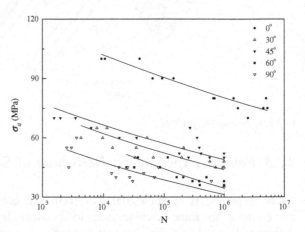

The results of the fatigue testing are presented in Figs. 2.4, 2.5, 2.6, and 2.7. Traditionally, constant amplitude fatigue data are plotted on the S–N plane. The number of cycles to failure is plotted on the abscissa, and the stress parameter on the ordinate. However, as mentioned in ASTM E739-91 (2004), the stress level must be considered as the independent variable, whereas the corresponding number of cycles to failure must be the dependent one. This is how the fatigue data have been used here for estimation of the derived S–N curve parameters.

Typical fitted lines of the form:

$$\sigma_a = \sigma_o N^{-\frac{1}{k}} \tag{2.4}$$

are also plotted together with the fatigue data. The coefficients σ_o and k are given in Table 2.2. The detailed fatigue data are given in Tables 2.3, 2.4, 2.5, and 2.6.

Conclusions concerning the fatigue behavior of the examined material can be derived from observation of these figures. Off-axis fatigue strengths follow the same trend as the corresponding static ones. Figure 2.4 for the reversed loading

Fig. 2.6 Fatigue data and
S–N curves for $R = 0.1$

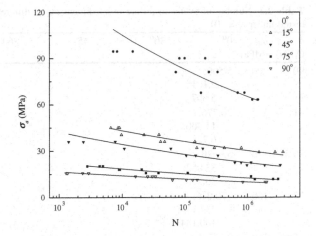

Fig. 2.7 Fatigue data and
S–N curves for $R = 0.5$

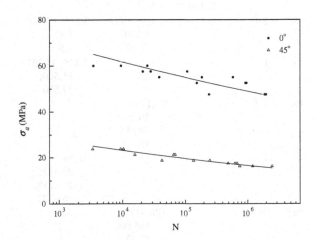

Table 2.2 Estimated model parameters for the on- and off-axis specimens. Equation of the form:
$\sigma_a = \sigma_o N^{-\frac{1}{k}}$

R-ratio Off-axis angle	10		−1		0.1		0.5	
	σ_o (MPa)	$1/k$	σ_o (MPa)	$1/k$	σ_o (MPa)	$1/k$	σ_o (MPa)	$1/k$
0°	137.13	0.0531	164.10	0.0517	276.16	0.1047	98.22	0.0505
15°					88.02	0.0775		
30°	159.91	0.1150	124.54	0.0755				
45°	115.68	0.0802	122.26	0.0662	80.27	0.0924	45.46	0.0732
60°	69.92	0.0842	152.67	0.1076				
75°					39.89	0.0838		
90°	35.22	0.0512	109.95	0.0910	27.98	0.0749		

Table 2.3 Number of cycles to failure of specimens cut at different on- and off-axis angles and tested under $R = 10$

σ_a (MPa)	σ_{max} (MPa)	0°	30°	45°	60°	90°
90	200	5,500; 2,161; 5,607				
81	180	20,776; 11,400; 6,728				
76.5	170	53,626; 33,052; 36,350				
72	160	770,046; 437,115; 100,184				
65.5	150	2,357,018; 431,315; 225,912				
58.5	130		6,819; 5,713; 8,500; 10,727	5,716; 8,972; 1,453; 6,519		
54	120			7,465; 15,257; 11,500; 57,500		
49.5	110		27,173; 26,292; 19,888; 15,329			
45	100			91,597; 25,317; 192,288; 161,427		
40.5	90		128,527; 195,000; 77,433; 142,397			
38.25	85			1,145,696; 1,338,602; 1,221,080		
33.75	75		850,000; 976,497; 1,050,000		11,675; 5,442; 10,545; 2,540	
29.25	65				16,943; 38,911; 26,841; 40,316	
24.75	55				150,000; 270,000; 317,000; 102,412	

(continued)

Table 2.3 (continued)

σ_a (MPa)	σ_{max} (MPa)	0°	30°	45°	60°	90°
22.5	50				710,316; 840,316; 896,316; 1,000,000 →	5,316; 2,158; 7,567; 10,595
20.25	45					93,315; 17,042; 92,141; 25,006
18.9	42					75,724; 896,052; 206,244; 658,432
16.65	37					1,111,693; 2,505,659; 1,554,429

(: → Run-out)

shows that the fatigue strength of the material at 45° is superior to the fatigue strength at 30°, which is in line with the static case and is attributed to the same reason, the presence of fibers along the 45° and −45° directions in the multidirectional laminate.

The slopes of the derived S–N curves have values that range between 0.05 and 0.12, in agreement with corresponding values in the literature for other GFRP systems, e.g. [25, 26]. For the on-axis specimens it is shown that the material is more sensitive to tensile fatigue loading patterns than to compressive ones.

2.2.4 Mean-Stress Effect

Comparison of the S–N curves for different R-ratios, see Figs. 2.8, 2.9, and 2.10, shows different behavior between the on- and off-axis specimens. For the on-axis specimens that fail due to the eventual failure of the fibers, it is shown (Fig. 2.8) that when fatigue life is plotted against the maximum cyclic stress, σ_{max}, the reversed loading is the worst case, with the tensile fatigue loading being the least critical, especially for low numbers of fatigue cycles, less than 800,000 in this case. This observation is in line with the quasi–static behavior of the material, which is stronger under tension than under compression. However, the accumulation of matrix cracks during loading that is more pronounced for the tensile loading (see following paragraphs) gradually reduces the fatigue strength of the material and therefore the S–N curve for $R = 0.1$ is steeper than the corresponding one for $R = 10$. The same comment (the reversed loading is the most critical) applies for the specimen cut at 45° from the multidirectional laminates, see Fig. 2.9, although

Table 2.4 Number of cycles to failure of specimens cut at different on- and off-axis angles and tested under $R = -1$

σ_a (MPa)	σ_{max} (MPa)	0°	30°	45°	60°	90°
100	100	10,700; 39,637; 9,350				
90	90	64,871; 93,498; 143,896				
80	80	670,275; 702,056; 1,446,527; 5,000,000→				
75	75	4,500,000→ 1,700,786; 5,269,524				
70	70			1,557; 3,500; 1,972		
65	65		7,820; 11,407	6,317; 270,633		
60	60		4,298; 52,316; 9,749	18,375; 357,155; 35,012		
55	55		14,098; 34,538; 212,856			3,641; 2,510; 3,000
52	52			179,000; 415,000; 1,000,000		
50	50		195,710; 33,149; 66,559	255,000; 580,995; 1,000,000→	36,774; 115,000	
48	48		581,997; 1,000,000			
45	45		956,933; 726,537; 1,000,000→		85,700; 32,000; 23,769	2,700; 18,000; 80,556; 25,000
42	42					29,850; 13,580; 46,120
40	40				209,560; 512,000; 161,000	26,916; 69,134; 16,027
38	38				388,000; 298,317; 1,000,000→	70,030; 30,206; 620,305
36	36				395,000; 1,000,000→ ; 1,007,000	
35	35					1,000,000; 1,000,000→

(:→Run-out)

in this case the S–N curve for tensile loading is significantly lower than that derived under $R = 10$. However, for specimens cut at 90°, the most critical loading pattern seems to be the tensile one, $R = 0.1$, see Fig. 2.10. The lack of fibers along the loading direction, which makes the material vulnerable to tensile loads, is the

Table 2.5 Number of cycles to failure of specimens cut at different on- and off-axis angles and tested under $R = 0.1$

σ_a (MPa)	σ_{max} (MPa)	0°	30°	45°	60°	90°
94.5	210	8,400; 7,284; 15,000				
90	200	82,000; 214,300; 99,783				
81	180	337,760; 245,995; 72,100				
67.5	150	898,645; 701,093; 182,123				
63	140	1,204,333; 1,464,000; 1,500,000				
45	100		9,125; 6,610; 8,654			
40.5	90		22,613; 9,834; 38,891			
36	80		41,330; 144,730; 47,894	7,819; 2,450; 1,420		
31.5	70			10,935; 59,821; 195,116		
31.95	71		298,233; 155,864; 432,283; 798,467			
29.25	65		3,828,947; 2,852,760; 1,549,331			
27	60			149,100; 105,000; 342,000		
22.5	50			793,000; 1,150,000; 632,089		
20.7	46			985,000; 1,915,000; 3,462,000		
20.25	45				4,419; 2,815; 4,965	
18	40				9,226; 20,715; 8,924	

(continued)

Table 2.5 (continued)

σ_a (MPa)	σ_{max} (MPa)	0°	30°	45°	60°	90°
15.75	35				111,152; 38,237; 24,111	1,296; 2,954; 2,370; 1,370
13.5	30				1,149,039; 354,521; 354,109	30,441; 25,581; 16,440; 34,181
11.7	26				1,325,554; 2,654,235; 3,200,000	
11.25	25					133,043; 104,913; 156,936; 63,520
9.9	22					839,958; 1,955,673; 2,075,673

Table 2.6 Number of cycles to failure of specimens cut at different on- and off-axis angles and tested under $R = 0.5$

σ_a (MPa)	σ_{max} (MPa)	0°	45°
60	240	25,210; 9,500; 3,500	
57.5	230	107,441; 28,461; 21,333	
55	220	38,970; 586,000; 185,694	
52.5	210	923,020; 956,833; 154,000	
47.5	190	1,996,000; 243,500; 1,900,000	
23.75	95		3,411; 9,370; 10,393
21.25	85		65,459; 15,858; 68,947
18.75	75		42,900; 135,872; 249,194
17.5	70		630,000; 493,345; 678,643
16.25	65		1,208,000; 750,000; 2,500,000→

(:→Run-out)

reason for this phenomenon. Although under quasi–static loading the 90° specimens exhibit almost identical strengths, the damage done to the material in the form of matrix cracking proved critical in the case of fatigue loading.

Similar conclusions are reported in other experimental works. In [25] it is mentioned that in multidirectional composite laminates with layers at 0° and 45°, with a stacking sequence of [0/±45/0]$_S$, the reversed loading is the most detrimental for the fatigue life since the strain-life (ε-N) curve for this case is lower

Fig. 2.8 S–N curves for different *R*-ratios. On-axis specimens

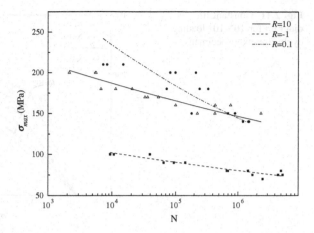

Fig. 2.9 S–N curves for different *R*-ratios. 45° off-axis specimens

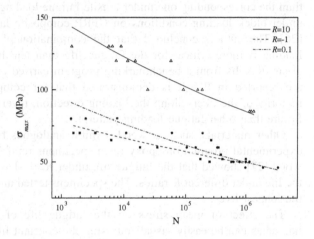

Fig. 2.10 S–N curves for different *R*-ratios. Specimens cut at 90°

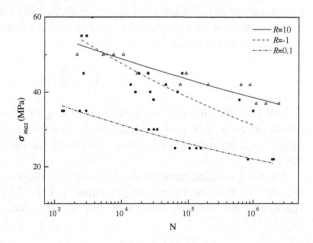

Fig. 2.11 Constant life diagram for 10^4–10^8 loading cycles, on-axis specimens

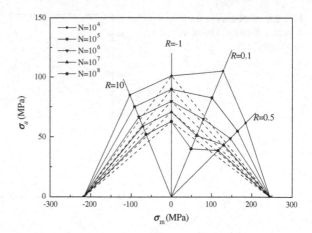

than the corresponding one under tensile fatigue loading of $R = 0.1$. Experiments under block loading conditions on CFRP composite laminates were reported in [26]. The authors concluded that the combination of tensile and compressive loading is more critical for the fatigue life than tensile or compressive loading alone. Results from a benchmarking program carried out by several laboratories are reported in [9]. It is demonstrated that for composite materials with the majority of the fibers along the loading direction, reversed loading causes earlier failure than other fatigue loading patterns.

Other materials, such as wood, exhibit analogous behavior. Results from an experimental program on epoxy resin specimens reinforced by different types of wood [27] showed that the fatigue life under $R = -1$ is the shortest compared to the life under different R-ratios. The specimens tested under tensile loads exhibited the longest lives.

The effect of mean stress on the fatigue life of the examined composite laminates can be easily visualized using the constant life diagrams. For the derivation of a constant life diagram, the fatigue data are normally plotted on the $(\sigma_m$–$\sigma_a)$-plane as radial lines emanating from the origin of the coordination system. Each line represents a single S–N curve at a given R-ratio and can be reproduced using the following equation:

$$\sigma_a = \left(\frac{1-R}{1+R}\right)\sigma_m \qquad (2.5)$$

Constant life lines can be derived by joining in a linear or non-linear way the points on each S–N curve (radial line) corresponding to the same number of cycles. Several methods have been proposed for the derivation of a CLD, see Chap. 4.

For the examined material the constant life diagrams for the on-axis and 45° and 90° off-axis specimens are presented in Figs. 2.11, 2.12, and 2.13. For the specimens cut on-axis and those cut at 45°, four S–N curves have been

Fig. 2.12 Constant life diagram for 10^4–10^8 loading cycles, 45° off-axis specimens

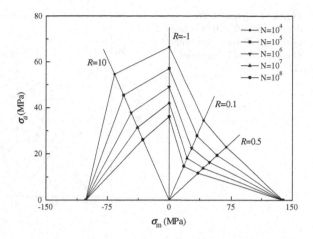

Fig. 2.13 Constant life diagram for 10^4–10^8 loading cycles, 90° off-axis specimens

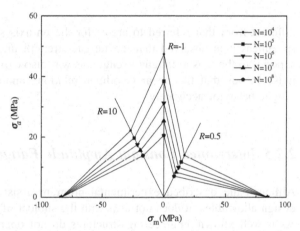

experimentally derived, under $R = 10, -1, 0.1$ and 0.5, while three, under $R = 10$, -1 and 0.1, are available for the 90° specimens.

It is obvious that for the on-axis specimens (Fig. 2.11) the constant life lines follow a Gerber line rather than the linear Goodman relationship. Therefore it is expected that the use of the Goodman diagram (shown by dashed lines in Fig. 2.11) as proposed in several design codes for structures made of composite materials, would lead to a very conservative design. Another interesting conclusion is that the material seems to be more fatigue resistant against compressive fatigue loads for the high cycle fatigue range, while the opposite holds true for small cycle numbers. This observation has also been mentioned by other researchers [14] for a similar GFRP material system, see Fig. 2.14. It is worth mentioning that in this case, even the linear Goodman relationship is less conservative compared to the experimental data for lives beyond 10^7.

The experimental results for the off-axis specimen are representative of the anisotropy of the examined material, since as presented the behavior is extremely

Fig. 2.14 Constant life
diagram for 10^4–10^8 loading
cycles. GL/P specimen with
fibers at 0/90/45 directions
[11]

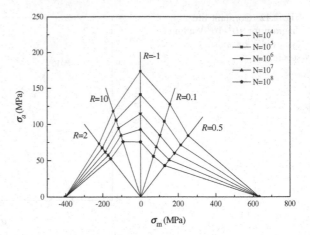

different from that referred to above for the on-axis specimens. For the off-axis specimens it is observed that the fatigue strength under compressive loading is superior to the tensile fatigue strength, as was also explained above. Furthermore, it is obvious that the linear Goodman relation cannot, in most cases, describe fatigue behavior accurately.

2.2.5 Interrupted Constant Amplitude Fatigue Loading

Although the described experimental program assists the derivation of fatigue design allowables, it does not deal with the subject of interrupted cyclic loading. As is well known, engineering structures do not operate continuously, but their operation is interrupted for several reasons, e.g. airplanes taking off, flying and landing, and then rest until the next flight, or wind turbines that stop when wind speed exceeds the limit of 25 m/s etc. It could thus be disputed whether or not the fatigue design allowables determined under continuous loading patterns are valid to accurately describe the fatigue behavior of composite materials and structures.

The influence of load interruption on fatigue life is examined here. Specimens cut from the same GFRP laminates at $0°$ direction were loaded at a single stress level under tension-tension ($R = 0.1$) fatigue loads until failure. Other specimens were loaded under the same conditions, but the loading was interrupted after a predetermined number of cycles. It was found that life was increased by a factor of 1.4 for this particular stress level.

Compared to experimental results from the aforementioned experimental program, it is shown that results from continuously loaded specimens fall within the experimental scatter of the S–N curve derived for $0°$ specimens, under $R = 0.1$. However, experimental results from specimens that were not loaded continuously could belong to another S–N curve, corresponding to longer fatigue life.

Table 2.7 Fatigue life of on-axis specimens under continuous and interrupted loading conditions

Specimen	Number of cycles to failure continuous loading	Number of cycles to failure Interrupted loading
1	51,316	242,546
2	42,163	203,629
3	150,319	64,050
4	86,197	296,903
5	400,316	268,345
6	57,460	209,279
7	550,316	187,598
8	87,280	63,397
9	58,963	379,900

The maximum applied load was maintained constant for all specimens at 11.1 kN, which corresponds to a maximum cyclic stress level of approximately $\sigma_{max} = 185$ MPa, according to each specimen's actual geometry. According to previous experimental data for the same material configuration, under the same test conditions, the allowable number of cycles for this cyclic stress level is approximately 200,000.

Half of the twenty specimens used in this short experimental program were loaded continuously until failure, and load reversals were recorded. The rest of the specimens were loaded for 5,000 cycles, and then left to rest for at least 1 h before resuming the test for another set of 5,000 cycles and so on until their failure. The total number of cycles until failure was also recorded for this type of test. Of the 20 fatigue tests conducted, 18 were considered valid, as one test from each group was rejected because of unacceptable failure mode; specimens failed in the tabs after a few cycles of loading. The fatigue life for the rest of the specimens is given in Table 2.7.

Assuming that the test data of each group follow a two-parameter Weibull distribution with a probability of survival:

$$P_S(N) = \exp\left[-\left(\frac{N}{N_o}\right)^{\alpha_f}\right]$$

(2.6)

The scale and shape parameters for the two distributions were estimated with MLE as $N_{\text{ocont}} = 170{,}790$ cycles and $\sigma_{f\text{cont}} = 1.08$, for continuous loading and $N_{\text{oint}} = 239{,}977$ cycles and $\sigma_{f\text{int}} = 2.37$ for interrupted loading. Subscript "cont" refers to continuous loading and subscript "int" denotes interrupted loading. The probability density function (PDF) of the Weibull distributions describing both groups of data are presented in Fig. 2.15 and characteristic values of cycles to failure are indicated for both distributions. Comparison of the characteristic numbers of cycles (the values of the scale parameters of the two distributions) leads to the conclusion that, as $N_{\text{oint}}/N_{\text{ocont}} = 1.41$, life is increased by 41% when loading is interrupted.

This comment continues to apply when experimental data are compared, i.e. at a reliability level of 50%. However, when a higher reliability level is desired, e.g.

Fig. 2.15 Probability density function of Weibull distributions describing fatigue data. Characteristic numbers of cycles are shown with vertical lines

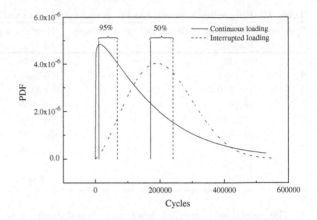

Fig. 2.16 Experimentally derived vs. assumed S-N curve

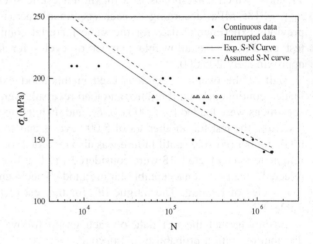

95%, conclusions are more obvious. In this case, for $P_S(N) = 95\%$, continuously loaded specimens will undergo $N_{cont(95\%)} = 10,914$ cycles prior to failure, while for specimens loaded under interrupted load patterns $N_{int(95\%)} = 68,492$ cycles. Consequently, the ratio of $N_{int(95\%)}$ over $N_{cont(95\%)}$ equals 6.28, meaning that specimen life could be considered as being approximately 630% longer at this reliability level.

It is therefore concluded that material life is longer if loading is interrupted several times during cycling. Therefore, fatigue design allowables, as determined to date, could be far too conservative, and, as demonstrated in Fig. 2.16, curves derived after non-continuous loading could possibly be used for design purposes. Here, the S–N curve determined after continuous testing under $R = 0.1$, at a frequency of 10 Hz, is compared to the hypothetical S–N curve, under interrupted loading at the same stress ratio and frequency conditions. Due to the lack of test data for the derivation of this second curve, it is supposed that it will have the same slope as that determined the conventional way, and it will be shifted to fit test data under interrupted loading at the maximum stress level of 185 MPa. Therefore, it will be a curve that corresponds to a life 1.4 times longer than that calculated using the experimentally determined S–N curve.

2.2.6 Variable Amplitude Loading

A total of 47 specimens were used for the variable amplitude loading investigation. Specimens cut at $0°$ and $30°$, $60°$ off-axis directions from the multidirectional laminate were examined experimentally. Two load spectra were used in applying variable amplitude loading on the specimens. For the application of the load, it was decided to keep the frequency constant but not the loading rate in order to obtain data under similar loading conditions to those for CA loading. Constant frequency results in higher loading rates, at high stress levels, while the opposite applies for lower loading levels. However, this is also the case in CA tests, during the determination of an S–N curve. Nevertheless, as has also been reported [28], there is no significant difference in the fatigue life of similar materials to those examined in this study if either constant frequency or constant loading rate is used.

The first applied load sequence is a modified version of a standardized spectrum (WISPERX) that is typically used in the wind turbine rotor blade industry for comparison purposes only in evaluating different materials or life prediction methods for example [29].

WISPERX spectrum is a short version of WISPER (WInd SPEctrum Reference), an irregular time series presented in 1988 [30, 31]. The development of this standardized spectrum was supervised by the IEA (International Energy Association) and a number of European industrial partners and research institutes active in the wind energy domain: FFA, Sweden, NLR and ECN, The Netherlands, Garrad Hassan and Partners and British Aerospace, UK, MAN Technologie GmbH and Germanischer Lloyd, Germany, RISO, Denmark, etc., participated in the working group. The spectrum was based on measurements of bending moment loadings at different sites and for different types of wind turbine rotor blades. A total number of nine different wind turbines with rotor diameters of between 11.7 and 100 m and rotor blades made of different materials such as metals, GFRP, wood and epoxy resins were considered. The WISPER standardized spectrum consists of 132,711 loading cycles, a high percentage of which are of low amplitude. Therefore a long period is needed for the failure of a specimen under this spectrum and for this reason the WISPERX spectrum was derived. It contains approximately $1/10^{th}$ of the WISPER loading cycles, while theoretically producing the same damage as its parent spectrum. The WISPERX spectrum comprises 12,831 loading cycles since all cycles with a range of below level 17 of WISPER have been removed.

The modified spectrum, henceforth denoted MWX, is in fact a shifted version of the original to produce only tensile loads. The lower positive stress level is the first level of the original time series in which level 25 is considered as zero stress

Fig. 2.17 MWX stress time series

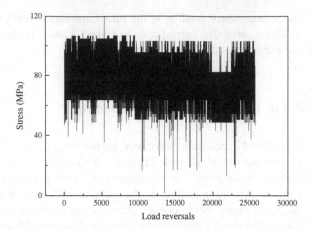

Fig. 2.18 EPET573 stress time series

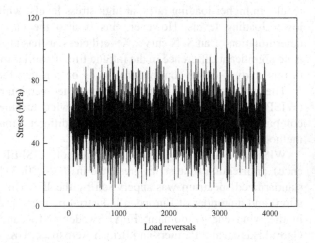

level. The same number of cycles as in the original, 12,831, is maintained. The stress time series for MWX of a maximum applied stress of 120 MPa is presented in Fig. 2.17. This load spectrum was applied on 30 specimens cut at 0°, 30° and 60° from the multidirectional laminate. Fifteen specimens at 0° were tested at three different maximum stress levels, 10 specimens cut at 30° also at three different maximum stress levels, and finally five specimens cut at 60° at two different stress levels. The test frequency was kept constant at 10 Hz and thus the time required from a peak to valley and vice versa was 0.05 s, irrespective of load range.

The second VA series used in the present experimental program was a more realistic spectrum, as it was derived through aeroelastic simulation [32] for a loading case of a 14 m rotor blade [33], and is representative of flap bending moment fluctuations on a cross section of the blade, located 2.4 m from the root. The load case definition [34] is for normal operating conditions at a wind speed of 21 m/s. It represents loading at the current cross-section for 10 min, and contains 3,893 loading reversals, or 1,946 cycles. When the time series is normalized with

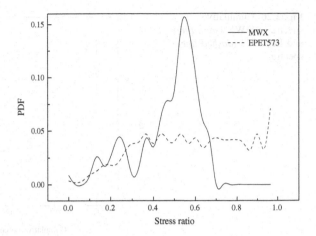

Fig. 2.19 Probability distributions of stress ratio values for MWX (*solid line*) and EPET573 (*dashed line*) spectra

respect to its minimum value and rounded off to integer numbers, it has a maximum value of 157, a minimum of 1, while the mean value is 71.47. The stress time series of this realistic spectrum, henceforth denoted EPET573, for a maximum applied stress of 120 MPa is presented in Fig. 2.18. A total of 17 specimens were tested under EPET573 as follows: eight specimens cut at $0°$, tested at two different maximum stress levels, six specimens cut at $30°$, tested at three different stress levels and three specimens cut at $60°$, tested at one stress level.

Both VA suites are composed of tensile cycles (T-T) of varying stress ratio, R. However, considerable differences are observed regarding the ranges and distribution of R–ratio values for each spectrum. For MWX, the range for R is [0.0516, 0.685] with a mean value of 0.486 and a standard deviation (S.D.) of 0.302, while the respective figures for EPET573 are [0.00345, 0.985], with a mean value of 0.603 and S.D. of 0.402. Probability density functions of R values for both spectra are shown in Fig. 2.19. It is apparent that EPET573 is composed of T-T cycles spanning the range with almost constant density whereas MWX has a much greater density of R values around its mean.

Observation of Figs. 2.17, 2.18 and 2.19 leads to the conclusion that the EPET573 spectrum is much more "irregular" than MWX, which can be considered as a sequence of blocks of almost constant loading patterns. WISPERX and consequently MWX is known by definition to contain mostly cycles of relatively high stress ranges, and has been created for Low Cycle Fatigue (LCF) conditions as it has been extracted from the WISPER spectrum by neglecting cycles of low amplitude. On the other hand, the EPET573 spectrum contains cycles of both high and low amplitudes, as directly derived from aeroelastic simulation. Apparently, even if these two load spectra were applied on a specimen for the same number of cycles at the same maximum applied stress level, MWX is expected to be more damaging than EPET573, which is mainly composed of low range cycles.

The above comments are more easily understood by observing the cumulative spectra of both time series, presented in Fig. 2.20. It is evident that EPET573 is more irregular than MWX, as its cumulative spectrum contains more steps. This is

Fig. 2.20 Cumulative spectra of MWX (*solid line*) and EPET573 (*dashed line*) spectra

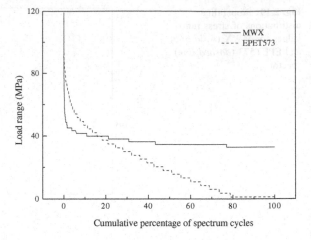

Cumulative percentage of spectrum cycles

Fig. 2.21 VA experimental results. Closed symbols: MWX, Open symbols: EPET573

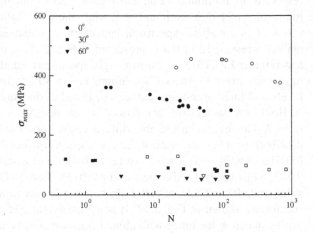

reflected in the 553 loading blocks that were used by the rainflow counting routine for the 1,946 counted cycles of the same range and mean values for EPET573, in contrast to only 114 loading blocks used to accommodate the 12,831 counted cycles of the MWX time series. Moreover, in Fig. 2.4 it is shown that MWX has no cycles with a range of less than 30% of its maximum range value, while more than 80% of the counted cycles of EPET573 have ranges below 30% and more than 60% counted cycles have ranges below 15% of the maximum range value.

For both test series, experiments were terminated upon specimen ultimate failure. Results are generally presented with respect to the number of passes of the applied irregular spectrum. Test data are presented in Fig. 2.21 for both MWX and EPET573 spectra, as number of passes vs. maximum applied stress of the VA time series. That is, the initial time series was multiplied each time point by point, by a factor, in order to achieve different levels of loading. This procedure resulted in time series having different mean, minimum and maximum values. The difference

Table 2.8 Experimental results from application of MWX and EPET573 spectra on on– and off–axis specimens

	σ_{max} (MPa)	Cycles to failure	Passes
0°			
wx1	296.48	296,976	23.15
wx2	300.00	330,360	25.75
wx3	296.30	289,466	22.56
wx4	291.88	586,408	45.70
wx5	299.07	396,222	30.88
wx6	295.16	398,734	31.08
wx7	323.18	151,038	11.77
wx8	366.52	7,505	0.58
wx9	319.41	192,850	15.03
wx10	315.58	301,588	23.50
wx11	281.07	671,640	52.35
wx12	283.83	1,671,388	130.26
wx13	360.00	25,510	1.99
wx14	336.62	110,262	8.59
wx15	360.00	30,033	2.34
ep1	378.21	1,120,744	575.92
ep2	375.16	1,326,439	681.62
ep3	451.92	215,502	110.74
ep4	453.71	185,078	95.11
ep5	455.52	66,584	34.22
ep6	426.94	40,260	20.69
ep7	453.71	189,208	97.23
ep8	439.75	512,938	263.59
30°			
wx16	79.40	1,012,000	78.87
wx17	90.10	202,008	15.74
wx18	86.64	339,616	26.47
wx19	83.82	495,924	38.65
wx20	114.42	16,172	1.26
wx21	119.35	6,563	0.51
wx22	81.55	1,050,007	81.83
wx23	79.24	1,452,918	113.23
wx24	115.37	17,875	1.39
wx25	86.06	965,482	75.25
ep9	129.03	42,416	21.83
ep10	127.04	15,185	7.82
ep11	99.64	219,642	113.04
ep12	98.18	419,736	216.02
ep13	83.95	1,610,035	828.63
ep14	83.63	911,328	469.03
60°			
wx26	62.71	41,966	3.27
wx27	60.98	145,026	11.30

(continued)

Table 2.8 (continued)

	σ_{max} (MPa)	Cycles to failure	Passes
wx28	53.18	992,635	77.36
wx29	55.86	378,151	29.47
wx30	52.67	617,283	48.11
ep15	60.77	218,809	112.61
ep16	64.60	100,380	51.66
ep17	60.30	224,303	115.44

Fig. 2.22 Comparisons of fatigue life of several GFPR materials under WISPERX spectrum

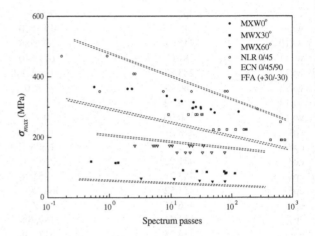

between the two fatigue spectra is well reflected in test results since for the same or even higher maximum stress, specimens tested under the EPET573 spectrum sustain the load for a greater number of passes.

The results of the application of MWX and EPET572 are also tabulated in Table 2.8. The maximum stress presented in Table 2.8 corresponds to the maximum peak of the VA spectrum, while the number of spectrum passes is calculated by dividing the number of cycles by the total number of cycles measured in the spectrum, i.e. 12,831 for MWX and 1,946 for EPT573.

A number of reports referring to experimental results exist [10, 23, 28, 35–37] from the application of the WISPERX spectrum on GFRP laminates. The examined materials were similar to the one examined here and it is therefore possible to compare their fatigue behavior. Such a comparison is shown in Fig. 2.22 for variable amplitude fatigue data collected in different laboratories.

As shown in Fig. 2.22, the variable amplitude fatigue data for the examined similar material can be classified in two scatter bands, depending on whether or not fibers exist along the loading direction. All on-axis specimens exhibit similar fatigue behavior, superior to the corresponding fatigue behavior of the 30° and 60° off-axis specimens.

2.3 Failure of Multidirectional Composite Laminates Under Fatigue Loading

The damage accumulation in composite materials is a multi-parametric phenomenon consisting of a number of different interacting mechanisms that finally lead to material failure. Quantification of the accumulated damage to the material, compared to the virgin version can provide a measure for the life of the examined material if the damage mechanisms can be recognized.

Many researchers have dealt with the measurement of the damage accumulation in fibrous composite materials loaded under static and/or fatigue loading and tried to develop failure criteria based on the measured damage metric. The idea is simple: as more damage is accumulated in the material, the closer it gets to failure. The rate of material deterioration depends on the developed failure mechanisms.

Numerous experimental programs have been performed in order to recognize the damage mechanisms that appear in composite materials and to model their development with fatigue loading. Masters and Reifsneider [38] reported the results of fatigue experiments under tensile loads on CFRP specimens. They measured damage development by interrupting loading periodically and making edge surface replicas. The surface replicas at different stages were then compared to an initial replica that was made, while the specimens were loaded under a 1.1 kN tensile load prior to fatigue testing. They could thus follow the development of the transverse cracks in each layer, and also the interlaminar cracks led to delamination of adjacent plies and ultimate failure. Ferry et al. [39] examined the bending and torsional fatigue behavior of translucent glass/epoxy specimens. A charge-coupled device (CCD) camera that was used to identify the damage development mechanisms recorded microcracks followed by ply delamination and finally fiber fracture. Talreja [3] presented and explained the different failure mechanisms that can develop in different types of fibrous composites and introduced a theory linking the size of each crack in the materials with the macroscopic variation of damage metrics, such as material stiffness.

The results of all the aforementioned theories, together with a number of others in the literature, for example [40–43] lead to the conclusion that the failure mode of a multidirectional laminate under fatigue loading depends mainly on the stacking sequence and loading type. In general, the failure mode under fatigue loads is similar to the failure mode under corresponding quasi-static loading patterns. However, more cracks are accumulated in the matrix of the material with fatigue loading and the type of cracks depends on the cyclic stress amplitude [38, 40]. The accumulation of these cracks results in additional stiffness degradation compared to the quasi-static loading, as reported for example in [40], where for the glass/epoxy examined material the stiffness degradation prior to fatigue failure is approximately 15%, more than double the corresponding stiffness degradation prior to quasi-static failure measured in the range between 3 and 7%.

The main types of damage that can develop during the fatigue failure of a multidirectional laminate can be classified as follows:

- Matrix cracks: Depending on the stress level, these cracks that initially appear in one of the layers can be transferred to adjacent plies under high stress levels, or restricted to the layer in which they initially appeared under low stress levels. These cracks can develop parallel or transverse to the loading [3]. In both cases, depending on the stress level, they can cause delamination or fiber fractures.
- Layer delamination: In many loading cases, the strain field that develops in a multidirectional laminate is such that it does not allow all layers to comply with the strain compatibility equations. In this case, interlaminar stresses develop and lead to delamination of adjacent layers. Subsequently, the layers act independently and not as part of the multidirectional laminate. As stated in [44], the independent layer usually exhibits lower strength than it would as part of the multidirectional laminate.
- Interface failure: The interface is a small region in between the matrix and the fibers where the two components are connected by mechanical and chemical bonds. Crack propagation in this region can be characterized as an interface failure.
- Fiber fracture: This is usually the last stage of damage accumulation in a fibrous composite material. The fibers are the main load-bearing component and their failure is linked to the ultimate material failure as the matrix cannot bear the applied loads without the presence of the fibers. In most cases however, before fiber failure occurs, the matrix is already damaged and unable to transfer any loads.

Although each of the aforementioned failure types can be independently recognized, in practice they act synergistically and appear simultaneously at different locations of the loaded material. Therefore, it is not easy to identify the failure mode of a multidirectional laminate and preferable to characterize one of the many failure modes that appear as the dominant one. The failure mode of a multidirectional laminate is a mixed one containing all the aforementioned types to some extent, depending mainly on stacking sequence and loading type.

These phenomena have been systematically identified and recorded by several means. Methods for the observation of damage initiation and development in composite materials are presented in [38, 41, 42, 45]. Almost all of the non-destructive techniques are also used for the recognition of damage patterns on the surface and also inside composite materials. The most common methods are the following:

- X-rays: With this method it is possible to identify a number of damage types throughout the material volume.
- Acoustic emission: It is known that any type of failure in a material increases the acoustic emission signal. The development of acoustic emission patterns for each failure type and for each material allows the characterization of the damage that causes this failure [45].
- Liquid penetrants and reflected light microscopy [41].
- Surface observation during fatigue loading or post mortem observation using a scanning electron microscope [42].

Fig. 2.23 Stiffness
degradation data for different
material configurations

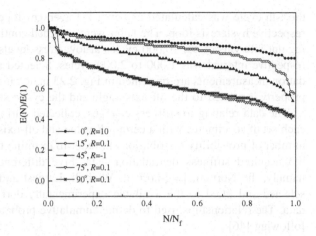

- Ultrasounds, C-scan and D-scan: Ultrasounds can be used for the recognition of damage patterns on the surface or even in the volume of a material.
- Surface replica: A method used in the past, based on the replication of the edge surfaces of the materials during fatigue loading in order to follow damage development, [38, 45].

   With all these methods now available, it is relatively easy to recognize the damage mechanisms that are triggered in a composite material during fatigue loading. However, it is difficult to model the interaction of the different mechanisms and predict the consequent material failure.

   The results of the experimental program presented in this chapter are consistent with others in the literature. All experimental results lead to the conclusion that the damage mechanism of each failure type for a given composite laminate is strongly dependent on the loading type (tensile or compressive loads) as well as on stress level. Under higher stress levels the failure is more abrupt with less damage accumulation in the material, as was also reported in [7] for a thermoplastic matrix APC-2 material with the stacking sequence $[45/0/-45/90]_{2S}$ loaded under $R = -1$. A result of this is lower decrease in stiffness that is observed when higher stress levels are present. In addition, it has been mentioned that the developed damage mechanisms are also dependent on the stacking sequence of the examined material. A high percentage of fibers along the load direction leads to less damage accumulation during fatigue life and subsequently less stiffness degradation. Photographs of failed specimens are shown below to support these comments.

2.3.1 Stiffness Degradation Measurements

Stiffness degradation was recorded during the constant amplitude fatigue experiment as described above in order to quantify the effect of the fiber orientation, the stress level and the R-ratio on the fatigue failure. The dynamic Young modulus at the i-th cycle was calculated as being the average slope of data points from the respective hysteresis loop. The latter were recorded continuously for 300 cycles at regular intervals of ca. 10,000 cycles at low stress levels. For higher stresses, the respective interval was 1,000 to 2,000 cycles. Selected results of stiffness degradation measurements are presented in Fig. 2.23 indicating that the variation of this property is related to the off-axis angle and the cyclic stress level.

The data relating to stiffness changes, collected during fatigue experiments on each set of specimens, with a certain R-value and off-axis angle θ, were fitted by a number of probability distributions in order to examine the stochastic behavior of the acquired stiffness degradation data. Five different statistical distributions, namely, the Normal, Log-normal, Weibull, Largest and Smallest element, were selected and fitted to the available experimentally derived stiffness degradation data. The relationships used to define cumulative probability distributions are the following [46]:

Normal:

$$F(x; \mu, \xi) = \int_{-\infty}^{x} \frac{1}{\xi\sqrt{2\pi}} \exp\left[-\frac{(z-\mu)^2}{2\xi^2}\right] dz \qquad (2.7)$$

Log-normal:

$$F(x; \mu, \xi) = \int_{-\infty}^{x} \frac{1}{\xi z\sqrt{2\pi}} \exp\left[-\frac{(\ln z - \mu)^2}{2\xi^2}\right] dz \qquad (2.8)$$

Weibull:

$$F(x; \eta, \xi) = 1 - \exp\left[-\left(\frac{x}{\xi}\right)^{\eta}\right], x \geq 0 \qquad (2.9)$$

Largest element:

$$F(x; \mu, \xi) = \exp\left[-\exp\left(-\frac{x-\mu}{\xi}\right)\right] \qquad (2.10)$$

Smallest element:

$$F(x; \mu, \xi) = 1 - \exp\left[-\exp\left(\frac{x-\mu}{\xi}\right)\right] \qquad (2.11)$$

It is noted that the symbol ξ is used here instead of the commonly used σ in textbooks to avoid confusing it with stress. The parameters of the distributions for each set of the examined data, based on maximum likelihood estimators, are given in Table 2.9.

The Kolmogorof-Smirnof (K-S) goodness of fit test was performed for each one of the hypotheses. The K-S test results, i.e. the D_N statistic and its probability, $P(D_N)$, for all the aforementioned distributions are given in Table 2.10. Values of

Table 2.9 Stiffness degradation: parameters of various statistical distributions

Distribution		Angle θ	0				15	30		45				60		75	90		
		R-ratio	10	−1	0.1	0.5	0.1	10	−1	10	−1	0.1	0.5	10	−1	0.1	10	−1	0.1
Normal	μ		0.926	0.941	0.734	0.870	0.897	0.951	0.849	0.932	0.833	0.917	0.889	0.945	0.799	0.685	0.968	0.692	0.926
	ζ		18.79	23.99	5.915	11.63	13.29	26.84	8.868	16.74	8.530	15.14	17.02	23.27	8.970	5.042	27.82	4.904	18.79
Log-normal	μ		0.902	0.919	0.684	0.834	0.862	0.932	0.805	0.904	0.788	0.884	0.866	0.945	0.761	0.632	0.946	0.638	0.902
	ζ		0.050	0.050	0.119	0.085	0.083	0.040	0.104	0.059	0.104	0.079	0.046	0.057	0.080	0.129	0.070	0.129	0.050
Weibull	ζ		0.925	0.942	0.738	0.872	0.900	0.950	0.852	0.931	0.835	0.919	0.886	0.945	0.797	0.690	0.978	0.697	0.925
	η		0.039	0.039	0.094	0.067	0.065	0.031	0.082	0.046	0.082	0.617	0.036	0.036	0.063	0.101	0.055	0.101	0.039
Largest element	μ		−0.104	−0.086	−0.396	−0.188	−0.154	−0.072	−0.226	−0.103	−0.248	−0.128	−0.146	−0.081	−0.278	−0.479	−0.059	−0.469	−0.104
	ζ		0.056	0.056	0.178	0.108	0.104	0.043	0.136	0.066	0.137	0.094	0.053	0.051	0.102	0.201	0.088	0.195	0.056
Smallest element	μ		0.880	0.897	0.630	0.795	0.824	0.914	0.758	0.877	0.741	0.848	0.845	0.903	0.725	0.574	0.914	0.580	0.880
	ζ		0.039	0.039	0.094	0.067	0.065	0.031	0.082	0.046	0.082	0.617	0.036	0.036	0.063	0.101	0.055	0.101	0.039

Table 2.10 K-S test results for the distributions with parameters given in Table 2.9

Distribution		Angle θ 0				15	30	45					60		75	90		
	R-ratio	10	−1	0.1	0.5	0.1	10	−1	10	−1	0.1	0.5	10	−1	0.1	10	−1	0.1
Normal	D_N	0.113	0.093	0.091	0.103	0.065	0.090	0.069	0.099	0.075	0.108	0.160	0.085	0.108	0.082	0.154	0.107	0.113
	$P(D_N)$	0.250	0.480	0.325	0.130	0.554	0.311	0.883	0.357	0.769	0.121	0.001	0.327	0.181	0.387	0.040	0.139	0.250
Log-normal	D_N	0.088	0.141	0.064	0.115	0.110	0.110	0.048	0.099	0.072	0.159	0.143	0.166	0.087	0.077	0.253	0.104	0.088
	$P(D_N)$	0.549	0.076	0.769	0.069	0.055	0.125	0.997	0.346	0.769	0.005	0.005	0.002	0.419	0.455	0.000	0.160	0.549
Weibull	D_N	0.100	0.103	0.126	0.102	0.049	0.112	0.104	0.113	0.113	0.096	0.213	0.101	0.157	0.144	0.237	0.174	0.100
	$P(D_N)$	0.386	0.343	0.063	0.143	0.872	0.115	0.425	0.206	0.234	0.214	0.000	0.158	0.013	0.012	0.002	0.018	0.386
Largest element	D_N	0.099	0.151	0.069	0.140	0.122	0.113	0.067	0.104	0.091	0.172	0.133	0.078	0.067	0.048	0.274	0.068	0.099
	$P(D_N)$	0.399	0.047	0.679	0.013	0.024	0.109	0.907	0.290	0.486	0.002	0.010	0.438	0.755	0.943	0.000	0.659	0.399
Smallest element	D_N	0.158	0.202	0.098	0.185	0.178	0.153	0.104	0.159	0.136	0.226	0.075	0.129	0.061	0.064	0.322	0.047	0.158
	$P(D_N)$	0.035	0.002	0.244	0.000	0.000	0.010	0.420	0.023	0.090	0.000	0.369	0.031	0.844	0.692	0.000	0.959	0.035

Fig. 2.24 Comparison of experimental and theoretical cumulative distributions of stiffness degradation $R = -1$, 30° off-axis

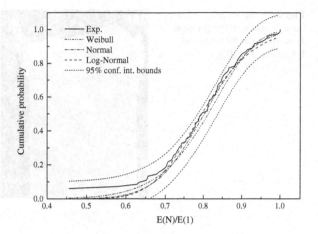

Fig. 2.25 Comparison of experimental and theoretical cumulative distributions of stiffness degradation $R = -1$, 0° on-axis

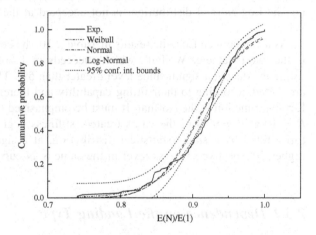

$P(D_N)$ greater or equal to 0.05 correspond to goodness of fit at a significance level of 5% or higher. The calculation for the K-S test were performed using the method described in [47].

As it is seen from the results in Table 2.10, stiffness degradation data, at a specific R-value and off-axis angle θ, can be modeled by a single statistical distribution for the whole range of stress levels considered for an S–N curve determination. It should be noted that for the cases studied here, the number of experimental points taken into account was at least 80.

An example of favorable and unfavorable comparisons between experimental data and theoretical distributions is given in Figs. 2.24 and 2.25, respectively for different material configurations.

The upper and lower 95% confidence interval bounds correspond to the less satisfactory distribution, which for Fig. 2.24 is the two-parameter Weibull and for

Fig. 2.26 Specimen loaded on-axis under **a** compression, $R = 10$, $\sigma_a = 80$ MPa and **b** tension, $R = 0.1$, $\sigma_a = 100$ MPa

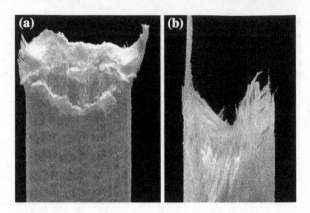

Fig. 2.25 is the Log-Normal. In this latter case, since the experimental sampling distribution intersects the 95% confidence bounds, the respective null hypothesis, i.e. the Log-Normal distribution, is not accepted at the significance level of 5% [48].

As it can be seen from the results in Table 2.10, the best performing distribution is the two-parameter Weibull, which succeeds in 14 of the 16 treated cases in fitting the data at a significance level greater than 5%. The statistical distributions are sorted according to their fitting capability and therefore, the second best performing function is the Normal. It must be emphasized that in all but one (that of $R = 10$ at $\theta = 90°$) of the cases treated, stiffness degradation data are satisfactorily fitted by a single statistical distribution, at a significance level of 5% or higher, irrespective of stress level in the same S–N curve.

2.3.2 Dependence on the Loading Type

Application of compressive loads results in less damage accumulation than application of only tensile loads. This is correlated to fractographic examinations indicating less matrix cracks in cases of compressive load dominance. A specimen tested under compression-compression, at stress amplitude of 80 MPa is shown in Fig. 2.26a while a specimen tested at 100 MPa, under tension-tension loading is shown respectively in Fig. 2.26b.

Differences in the failed surfaces are obvious. The C-C loaded specimens failed after local damage accumulation, close to the ultimate fracture region as can be concluded after inspection of all the failed specimens under these loading conditions. In contrast, when the specimens were tested under tensile loads, detectable damage is homogeneously distributed throughout the specimen, as fiber splitting and matrix cracking are present, even in regions far from the separation area. Mean values of measured stiffness degradation data, prior to failure, from all specimens tested at specific stress levels and R-ratios follow this trend. As shown in Fig. 2.27,

Fig. 2.27 Stiffness
degradation vs. loading type
and off-axis angle

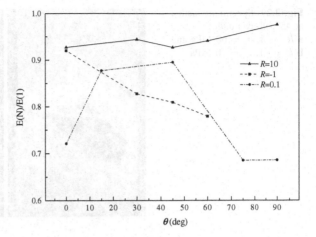

Fig. 2.28 Stiffness
variations during fatigue life.
Specimens cut at 45°, tested
under $R = -1$

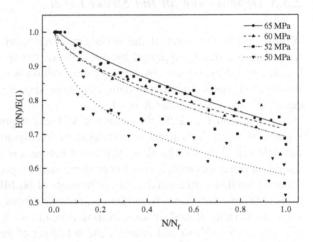

tension-tension and reversed fatigue loading patterns, both containing tensile
loading components, produce significantly higher stiffness degradation than the
compression-compression fatigue, independent of off-axis angle. It is concluded
that tensile loads promote damage accumulation during fatigue life, unlike specimens tested under C-C loads whose elastic modulus seems almost unaffected
during lifetime and changes only just before failure. The behavior of the C-C
loaded specimens can be affected by the used antibuckling device, shown in
Fig. 2.3. The aluminum plates of the antibuckling guide do not allow the specimen
to suffer buckling, but also restrict any out-of-plane deflection that might occur due
to matrix failures. Therefore, it is conjectured that the absence of matrix cracks
during compressive loading may be an artifact due to the use of the antibuckling
guide. It should be mentioned that these failure patterns are representative of all
specimens tested under T-T or C-C loading.

Fig. 2.29 Specimens loaded at 45° off-axis under **a** $R = -1$, $\sigma_a = 60$ MPa and **b** $R = -1$, $\sigma_a = 70$ MPa

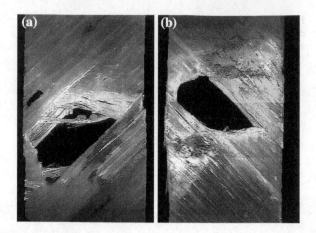

2.3.3 Dependence on the Stress Level

With regard to the effect of the stress amplitude level on the observed modulus reduction, the trend, supported by c.a. 75% of the specimens tested, is that for higher applied cyclic stresses, stiffness degradation is lower. This is derived from measured stiffness degradation data, like those shown in Fig. 2.28 for specimens cut at 45° and tested under $R = -1$.

As shown in Fig. 2.28, the exhibited stiffness degradation follows the afore-mentioned trend. This trend is correlated to fractographic examinations, see for example Fig. 2.29, in which, as mentioned before, the observed higher degree of matrix damage accumulation at lower stress states (specimen on left tested under reversed loading with a maximum cyclic stress of 60 MPa) can be associated with higher stiffness reduction. Nevertheless, failed surfaces of specimens tested at the same stress ratio, R, and off-axis loading orientation, θ, are very similar and it is not easy to distinguish and quantify the net effect of the cyclic stress level based only on their observation.

It must be emphasized however, that when differences in the failure patterns are clearly distinguishable, greater stiffness degradation can occur in the specimens that are more damaged. For example, under reversed loading, $R = -1$, the specimen shown in Fig. 2.29a, tested at the lower stress level, is more damaged than that in Fig. 2.29b, loaded at higher cyclic stresses. More matrix cracks were accumulated in the material in Fig. 2.28a and can be distinguished as white lines parallel to the 45° fibers of the outer layer. Thus, the stiffness degradation exhibited by the specimen in Fig. 2.29a is higher than the corresponding variation in the stiffness in Fig. 2.29b, as shown in Fig. 2.30.

An exception is presented in Fig. 2.31, where, as shown, specimens fatigued on-axis under T-T, $R = 0.1$, do not follow the general trend, i.e. "the higher the stress level, the lower the stiffness degradation". However, it is obvious that the specimen in Fig. 2.31a is more damaged, with evident matrix deterioration before failure although tested under a higher stress level than in Fig. 2.31b. The measured

Fig. 2.30 Stiffness
degradation of specimens
shown in Fig. 2.23

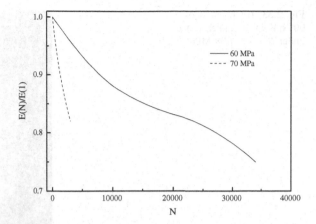

Fig. 2.31 On-axis loaded
specimens, **a** $R = 0.1$
$\sigma_a = 80$ MPa, **b** $R = 0.1$,
$\sigma_a = 65$ MPa

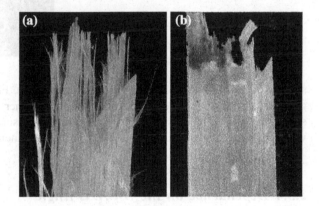

Fig. 2.32 Stiffness
degradation measurements
for individual specimens cut
on-axis

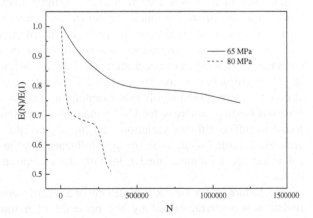

Fig. 2.33 Lateral view of 90° off-axis specimen tested under $R = -1$ at 38 MPa

stiffness degradation for these two specimens (see Fig. 2.32) contradicts the general trend, since the specimen tested under the higher stress level exhibits more stiffness degradation. Nevertheless, the stiffness measurements are in agreement with the observed damage accumulation in the specimens.

2.3.4 Dependence on the Off-Axis Angle

The combined action of in-plane stress tensor components seems to affect stiffness degradation in all cases except the C-C fatigue loading, where, as explained before, the use of the antibuckling device may have masked these effects. For higher off-axis angle specimens, an increased contribution of the transverse normal and the shear stress components to the stress tensor can be calculated and is reflected in higher stiffness degradation values, as shown in Fig. 2.27, where mean values of stiffness degradation data of all specimens, prior to failure, are plotted versus the corresponding off-axis orientation. This is unconditionally valid for reversed loading, whereas for C-C conditions, combined in-plane stress was not found to affect stiffness variations during fatigue life. As far as T-T loading is concerned, results seem to be strongly influenced by the combined in-plane stress, following the aforementioned rule with the exception of specimens cut at 0°, on-axis.

The failure of the off-axis specimens was matrix-dominated, as damage accumulation was characterized by the presence of a multitude of matrix cracks.

Fig. 2.34 Compression-compression fatigue loading **a** on-axis specimen at $\sigma_a = 70$ MPa, **b** 45° off-axis specimen at $\sigma_a = 40$ MPa

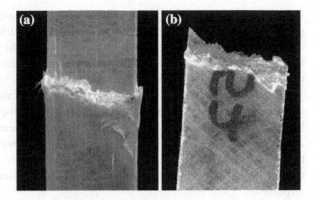

Specimens failed when matrix crack density reached a critical limit. The presence of a large number of matrix cracks was visually verified for off-axis specimens (see for example Fig. 2.33, for a 90° off-axis specimen tested under $R = -1$ with a maximum stress of 38 MPa).

On the other hand, fiber-failure mode was the dominant failure mechanism for on-axis specimens, regardless of stress ratio. Therefore, failure of specimens axially loaded along the main fiber direction was more abrupt and fewer stiffness variations occurred during the fatigue life. In contrast to this rule, specimens tested under $R = 10$ failed in the same mode, without any visible indication of crack accumulation away from the failure area, irrespective of the on- or off-axis angle or cyclic stress amplitude, see Fig. 2.34.

Therefore, the observed stiffness degradation data for these specimens was not found to depend on the off-axis angle or applied cyclic stress amplitude since, as measurements showed, it was almost unaffected during lifetime and changed abruptly just prior to failure.

2.4 Conclusions

The behavior of specimens cut from a multidirectional laminate with the stacking sequence $[0/(\pm45)_2/0]_T$, typical of that used in several lightweight structures, loaded under quasi-static and fatigue loading patterns has been examined in this chapter. The derived database contains 324 fatigue data and 31 quasi-static data.

The experimental results proved that the fatigue strength of the examined material follows the same trend as its static strength. It is higher in the case of on-axis specimens and diminishes with the off-axis angle. The unexpected behavior around the off-axis angle of 45° can be attributed to the stacking sequence of the laminates. The fibers along the ±45° directions are responsible for this increase in strength when specimens cut at this angle are tested.

The mean stress has been proved to have a significant effect on the fatigue life of the examined specimens, independent of the off-axis angle. As can be concluded from the constant life diagrams for the examined on-axis specimens, a Gerber parabola seems more appropriate than the linear Goodman representation. On the other hand, the material seems to be more sensitive to tensile loads only for the low cycle regime. However, experimental results obtained from off-axis specimens proved more resistant against compressive loads.

Another interesting conclusion concerns the interrupted loading results presented. It has been proved experimentally that the load interruptions were beneficial for the fatigue life of the material. Therefore the fatigue failure allowables derived from fatigue tests at high frequencies can be considered as being conservative when they apply to structures that do not operate continuously, such as wind turbine rotor blades, helicopter rotor blades and airplanes.

Fractographic examination of specimens after failure led to useful conclusions regarding damage development and the way in which this damage is related to the loading and the stacking sequence of the examined laminate.

The failure of a multidirectional laminate is a result of the combined effect of a number of interacting phenomena: matrix failure, layer debonding and finally fiber rapture can be observed.

The failure is also related to the loading pattern and is generally different for compressive and tensile loading. However, even when the loading type is the same, the failure mode may be different depending on the stress level. Lower stress levels cause more damage in the material before failure which consequently leads to more stiffness degradation.

The matrix failure and layer delaminations were observed in all failed specimens, especially those loaded under low stress levels. On the other hand, fiber failure is characteristic of higher stress levels and restricted to on-axis specimens, with the exception of the C-C loaded specimens. However, for the examined multidirectional laminates it is not possible to discriminate between pure fiber and matrix failure. Generally speaking, a mixed-mode was observed. Matrix cracks (their number depending on material configuration and loading conditions) developed, followed by layer delaminations and eventually fiber failure.

Investigation of the damage mechanisms shows the way in which damage is accumulated in the material. However, the complexity of these mechanisms and the development of several of them simultaneously, in contrast to metals for example where a single crack is created and leads to material failure, makes it very difficult, if not impossible, to develop a theoretical model that can be based on the fracture mechanics and accurately predict the fatigue behavior of this kind of material. Existing studies, e.g. [3, 49], have proved that the correlation of actual damage mechanisms to easily measured macroscopic quantities for the fatigue life estimation is a very difficult task that can easily lead to erroneous results. It is therefore considered more sensible, especially for design purposes, to adopt fatigue theories based on macroscopic quantities, e.g. cyclic stress and stiffness, for life prediction. Furthermore, the experimental results presented in this chapter for a given glass polyester composite laminate are very similar to others found in the literature for analogous laminates, used in similar engineering applications. This observation is encouraging for the development of a macroscopic fatigue failure methodology capable of estimating the fatigue life of composite materials independent of the fiber or matrix type and laminate configuration.

References

1. K.L. Reifsneider, K.N. Lauraitis (eds.), *Fatigue of filamentary composite materials, ASTM STP 636* (American Society for Testing and Materials, Philadelphia, 1977)
2. K.N. Lauraitis (ed.), *Fatigue of Fibrous Composite Materials, ASTM STP 723* (American Society for Testing and Materials, 1981
3. R. Talreja, Fatigue of Composite Materials, Technomic, (1987)
4. P.A. Lagace (ed.), *Composite Material: Fatigue and Fracture, ASTM STP 1012*, vol. 2, (American Society for Testing and Materials, Philadelphia, 1989)
5. K.L. Reifsnider (ed.), *Fatigue of Composite Materials*, Composite Materials Series 4, (Elsevier, Amsterdam, 1991)
6. T.K. O'Brien, Towards a damage tolerance philosophy for composite materials and structures, in *Composite Materials: Testing and Design ASTM STP 1059*, vol. 9, ed. by S.P. Garbo (American Society for Testing and Materials, Philadelphia, 1990)
7. S.G. Pantelakis, T.P. Philippidis, T.B. Kermanidis, Damage accumulation in thermoplastic laminates subjected to reversed cyclic loading. in *High Technology Composites in Modern Applications*, ed. by S.A. Paipetis, A.G. Yioutsos. (University of Patras, Patras 1995) pp. 156–1641
8. C.W. Kensche (ed.), *Fatigue of Materials and Components for Wind Turbine Rotor Blades*, European Commission, Directorate-General XII, Science, Research and Development, EUR 16684, (1996)
9. R.M. Mayer, *Design of Composite Structures Against Fatigue: Applications to Wind Turbine Blades*. (Antony Rowe Ltd., Chippenham, Wiltshire, 1996)
10. B.J. de Smet, P.W. Bach, Database fact: fatigue of composites for wind turbines, ECN-C-94-045, (1994)
11. J.F. Mandell, D.D. Samborsky, DOE/MSU Composite Material Fatigue Database. Sandia National Laboratories, SAND97-3002, online via www.sandia.gov/wind, v. 18, 21st March Updated (2008)
12. D.R.V. van Delft, H.D. Rink, P.A. Joosse, P.W. Bach, Fatigue behaviour of fibreglass wind turbine blade material at the very high cycle range. in *European Wind Energy Conference Proceedings*, vol. 1. (Thessaloniki, Greece, 1994) pp. 379–384
13. A.T. Echtermeyer, Fatigue of glass reinforced composites described by one standard fatigue lifetime curve. In *European Wind Energy Conference Proceedings*, vol. 1 (Thessaloniki, Greece, 1994) pp. 391–396
14. P.A. Joosse, D.R.V. van Delft, P.W. Bach, Fatigue design curves compared to test data of fibreglass blade material, in *European Wind Energy Conference Proceedings*, vol. 3, (Thessaloniki, Greece, 1994) pp. 720–726
15. C.W. Kensche, Lifetime of gl-ep rotor blade material under impact and moisture, 3rd Symposium on Wind Turbine Fatigue Proceedings, Petten, Holland: IEA, April 21–22, (1994) pp. 137–143
16. D.R.V. van Delft, G.D. de Winkel, P.A. Joosse, Fatigue behavior of fiberglas wind turbine blade material under variable loading, *4th Symposium on Wind Turbine Fatigue Proceedings*, Stuttgart, Germany: IEA, Feb 1–2, (1996) pp. 75–80
17. C.W. Kensche. Which slope for GL-Ep fatigue curve? *4th Symposium on Wind Turbine Fatigue Proceedings*, Stuttgart, Germany: IEA, Feb 1–2, (1996) pp. 81–85

18. M.J. Owen, G.R. Griffiths, Evaluation of biaxial stress failure surfaces for a glass fabric reinforced polyester resin under static and fatigue loading. J. Mater. Sci. **13**(7), 1521–1537 (1978)
19. Toru. Fujii, Fan. Lin, Fatigue behavior of a plain-woven glass fabric laminate under tension/ torsion biaxial loading. J. Compos. Mater. **29**(5), 573–590 (1995)
20. T.P. Philippidis, A.P. Vassilopoulos, Fatigue strength prediction under multiaxial stress. J. Compos. Mater. **33**(17), 1578–1599 (1999)
21. T.P. Philippidis, A.P. Vassilopoulos, Fatigue of composite laminates under off-axis loading. Int. J. Fatigue **21**(3), 253–262 (1999)
22. T.P. Philippidis, A.P. Vassilopoulos, Fatigue design allowables of grp laminates based on stiffness degradation measurements. Compos. Sci. Technol. **60**(15), 2819–2828 (2000)
23. S.I. Andersen, P.W. Bach, W.J.A. Bonee, C.W. Kensche, H Lilholt, A. Lystrup, W. Sys, *Fatigue of Materials and Components for Wind Turbine Rotor Blades*, ed. by C.W. Kensce, Directorate-General XII, Science, Research and Development, EU-16684 EN, (1996)
24. S.W. Tsai, H.T. Hahn, Introduction to Composite Materials. Technomic (1980)
25. P.W. Bach, Glass and hybrid fibre performance. in *Design of Composite Structures Against Fatigue. Applications to Wind Turbine Blades*, ed. by R. M. Mayer (Antony Rowe Ltd., Chippenham, Wiltshire, 1996)
26. B. Harris, N. Gathercole, H. Reiter, T. Adam, Fatigue of carbon-fibre-reinforced plastics under block-loading conditions. Compos. Part A Appl. Sci. **28**(4), 327–337 (1997)
27. M. Ansell, I. Bond, P. Bonfield, C. Hacker, Fatigue properties of wood composites. in *Design of Composite Structures Against Fatigue. Applications to Wind Turbine Blades*, ed. by R.M. Mayer (Antony Rowe Ltd., Chippenham, Wiltshire, 1996)
28. C.W. Kensche, *GFRP Fatigue Data for Certification. European Wind Energy Conference Proceedings*, vol. I, (Thessaloniki, Greece, 1994) pp. 738–742
29. T.P. Philippidis, A.P. Vassilopoulos, Life prediction methodology for GFRP laminates under spectrum loading. Compos Part A–Appl S **35**(6), 657–666 (2004)
30. A.A. Ten Have, Wisper: Introducing variable-amplitude loading in wind turbine research. *The 10th BWEA Conference*, London, UK, Mar 23–25 (1988)
31. A.A. Ten Have. Wisper: A standardized fatigue load sequence for HAWT-blades. European Community Wind Energy Conference proceedings, Henring Denmark, June 6–10, (1988) pp. 448–452
32. V.A. Riziotis, S.G. Voutsinas, Fatigue loads on wind turbines of different control strategies operating in complex terrain. J. Wind. Eng. Ind. Aerod. **85**(3), 211–240 (2000)
33. T.P. Philippidis, D.J. Lekou, A.P. Vassilopoulos, EPET II #573, University of Patras, First Semester Progress Report, (in Greek) (1997)
34. Draft IEC 61400-1, Ed.2 (88/98/FDIS): Wind turbine generator systems–Part 1: Safety requirements, (1998)
35. P.W. Bach, P.A. Joose, D.R.V. van Delft. Fatigue lifetime of glass/polyester laminates for wind turbines. *In the European Wind Energy Conference Proceedings*, vol. I, (Thessaloniki, Greece, 1994) pp. 94–99
36. M. Poppen, P. Bach, Influence of spectral loading. in *Design of Composite Structures Against Fatigue. Applications to Wind Turbine Blades*, ed. by R.M. Mayer (1996)
37. R.P.L. Nijssen, OptiDAT–fatigue of wind turbine materials database, 2006. http://www. kc-wmc.nl/optimat_blades/index.htm
38. J.E. Masters, K.L. Reifsneider, An investigation of cumulative damage development in quasi-isotropic graphite/epoxy laminates. in *Damage in composite Materials, ASTM STP 775*, ed. by K.L. Reifsneider, American Society for Testing and Materials (1982) pp. 40–62
39. L. Ferry, D. Rerreux, D. Varchon, N. Sicot, Fatigue behaviour of composite bars subjected to bending and torsion. Compos. Sci. Technol. **59**(4), 575–582 (1999)
40. G. Caprino, G. Giorleo, Fatigue lifetime of glass fabric/epoxy composites. Compos. Part A Appl. Sci. **30**(3), 299–304 (1999)

41. F. Gao, L. Boniface, S.L. Ogin, P.A. Smith, R.P. Greaves, Damage accumulation in woven fabric laminates under tensile loading: Part I. Observations of damage accumulation. Compos. Sci. Technol. **59**(1), 123–136 (1999)
42. E.K. Gamstedt, L.A. Berglund, T. Peijs, Fatigue mechanisms in unidirectional glass-fibre-reinforced polypropylene. Compos. Sci. Technol. **59**(5), 759–768 (1999)
43. M.M. Ratwani, H.P. Kan, Effect of stacking sequence on damage propagation and failure modes in composite laminates. in *Damage in composite materials, ASTM STP 775*, ed. by K.L. Reifsneider, (American Society fot Testing and Materials, 1982) pp. 40–62
44. A. Rotem, Prediction of laminate failure with the rotem failure criterion. Compos. Sci. Technol. **58**(7), 1083–1094 (1998)
45. T.P. Philippidis, V.N. Nikolaidis, J.G. kolaxis, Unsupervised pattern recognition techniques for the prediction of composite failure. J. Acoust. Emission. **17**(1–2), 69–81 (1999)
46. G.J. Hahn, S.S. Shapiro, *Statistical Models in Engineering* (Wiley, New York, 1994)
47. W.H. Press, S.A. Teukolsky, W.T. Vetterling, B.P. Flannery, in *Numerical Recipes in Fortran. The Art of Scientific Computing*, 2nd edn. (Cambridge University Press, Cambridge, MA, 1994)
48. F.J. Massey, The Kolmogorov-Smirnof test for goodness of fit. J. Am. Stat. Assoc. **46**, 68–78 (1951)
49. G.P. Sendeckyj, Life prediction for resin-matrix composite materials. in *Fatigue of Composite Materials*, Composite Materials Series 4, ed. by K.L. Reifsnider (Elsevier, Amsterdam, 1991)

Chapter 3
Statistical Analysis of Fatigue Data

3.1 Introduction

The fatigue design of a structural application is generally based on full-scale fatigue test results. However, due to time and cost constraints, the replication of this kind of experiment is always limited. Therefore, in order to increase design reliability, experimental programs are performed on specimens in parallel and the supplemented experimental results are analyzed. The behavior of the examined material must be modeled. However, mathematical models expressed by deterministic equations, which can describe the behavior of any material system, cannot easily be developed due to uncertainty regarding several factors such as the scatter of the examined population and the unpredictable parameter relationship.

The experimentally derived fatigue data for fiber-reinforced composite materials present high scatter that can reach the order of one decade of life. This is mainly owing to the nature of the reinforcements (fibers and fabrics) and the fabrication methods that, often, cause a large number of defects in the specimen. As a result, the S–N curve that is derived based on the fitting of the mean or the median lifetime versus the corresponding stress level is inadequate, since it provides no information concerning modeling reliability. This deterministic S–N curve equation yields an estimate of the mean time to failure as a function of the stress level. Such a procedure however, fails to take into account the large variation in the time to failure at a given stress level. Thus, when the time to failure has been reached, half of the samples have failed. To overcome this, several methods have been presented in the literature for the statistical analysis of composite material fatigue data. The objective is to derive S–N curves that correspond to high reliability levels in the range above 90% and conform with design codes. The derivation of models for the description of the material's fatigue behavior with some statistical significance requires the production of a large number of fatigue data per stress level in order to measure the distribution of the time to failure. However, several models have been presented in the past to overcome this problem and provide reliability-based S–N curves derived from limited datasets.

A. P. Vassilopoulos and T. Keller, *Fatigue of Fiber-reinforced Composites*,
Engineering Materials and Processes, DOI: 10.1007/978-1-84996-181-3_3,
© Springer-Verlag London Limited 2011

The simplest way to derive statistically meaningful S–N curves is to assume a normal distribution of the lifetime per stress level and estimate the characteristic number of cycles to failure for each stress level, for a given reliability level and coefficient of variation. The S–N curve is derived by fitting the characteristic number of cycles per stress level to the corresponding cyclic stress value.

Three other methods from the literature will be discussed in this chapter. The first one is the standard practice for statistical analysis of linear or linearized stress-life (S–N) and strain-life (ε–N) fatigue data, as proposed by ASTM [1]. This practice covers only S–N or ε–N relationships that may be reasonably approximated by a straight line (on appropriate coordinates) for a specific interval of stress or strain.

The second method, proposed by Whitney [2, 3], is based on two assumptions: (a) the S–N curve is of power form and (b) the lifetime per stress level follows a two-parameter Weibull distribution. This method needs an adequate number of replications per stress level (minimum 2) in order to estimate the parameters of the Weibull distribution. The formulation also takes into account the run-outs (the experimental results recorded before the occurrence of fatigue failure).

The third method was introduced by Sendeckyj [4]. It is based on three main assumptions: (a) the S–N curve is of deterministic nature but can be of any form, (b) the static strength of the material is related to the fatigue strength, i.e.the stronger specimen in fatigue loading would also be the stronger in static loading. Although this assumption cannot be experimentally verified (a specimen that fails under static loading cannot be loaded again under fatigue in order to verify the assumption), it is used by Sendeckyj to correlate static and fatigue strength and assign a static strength value to each fatigue experiment. Therefore the method can be applied to fatigue data where no replication per stress level is available, (c) The static strength data follows a two-parameter Weibull distribution. The method also takes into account the run-outs by considering them during the conversion of fatigue to static strength.

The aforementioned methods will be presented in this chapter and applied to several composite materials' constant amplitude fatigue data to evaluate their performance and compare the resulting reliability-based S–N curves.

3.2 Methods for the Statistical Analysis of Stress-Life Fatigue Data

3.2.1 Normal Lifetime Distribution

This is a simplified method that is based on the following assumptions:

(a) The probability distribution function of the cycles to failure per stress level is assumed to be normal,
(b) The value of the coefficient of variation equals 15%,

(c) The confidence interval for the lowest average of the fatigue data population is 95%.

The characteristic value of the cycles to failure per stress level, corresponding to 15% coefficient of variation and 95% confidence, can then be calculated by:

$$R_{ki}(5\%, 95\%, 15\%, m_i) = \bar{N}_i \left(1 - 0.15 \left(1.645 + \frac{1.645}{\sqrt{m_i}} \right) \right) \qquad (3.1)$$

for a dataset of m_i specimens tested at the corresponding stress level and exhibiting an average fatigue life of $\bar{N}_i$. The S–N curve for 95% reliability level can be derived based on the fitting of the characteristic number of cycles per stress level versus the corresponding stress level, considering the stress parameter as the independent variable. A fitting curve of any type can be used for the simulation of the S–N curve. If the power curve type is selected, the S–N curves of the material based on this method will be given by:

$$\sigma = \sigma_o R_k^{\left(-\frac{1}{k}\right)} \qquad (3.2)$$

where σ denotes the cyclic stress parameter under which the material exhibits life that can be fitted by a normal distribution with characteristic value, R_k. The slope of the S–N curve, $1/k$ and the Y-intercept, σ_o, are the model parameters that must be estimated by the fitting.

3.2.2   ASTM Standard Practice

The ASTM E739-91 (reapproved in 2004) standard practice [1] is applied for the derivation of the material's S–N or ε–N relationships that can be considered linear when the fatigue data are plotted on appropriate coordinates. The practice is not recommended for extrapolation outside the region of the available experimental data or for estimation of the fatigue life at a specific stress or strain amplitude for probability values below 0.05 or above 0.95 [5] The practice is not applicable for databases containing run-outs (no failure at a specified number of loading cycles).

Linear or linearized S–N relationships are considered:

$$\begin{aligned} \log(N) &= A + B\sigma \text{ or} \\ \log(N) &= A + B\log(\sigma) \end{aligned} \qquad (3.3)$$

In which refers to stress or strain parameter and N to the corresponding number of cycles to failure, while A and B are the model parameters that will be determined after the application of the process on the available fatigue data. The fatigue life is considered as the dependent variable whereas the stress or strain is considered as the independent variable for the fitting.

The distribution of the fatigue life is unknown. However, according to the ASTM practice, it is assumed that the logarithms of the fatigue cycles are normally distributed, which means that the fatigue life is log-normally distributed. In addition, it is assumed that the variance of log life is constant over the entire range of the independent variable, σ, i.e., the scatter in log life is the same for low and high stress levels. The dependent variable, $\log(N)$, will be denoted by Y, while the independent variable σ, or $\log(\sigma)$, will be represented by X in the following for simplicity. Therefore, Eq. 3.3 can be rewritten as:

$$Y = A + BX \tag{3.4}$$

If all requirements are met, the maximum likelihood estimators for A and B can be calculated by:

$$\hat{A} = \bar{Y} - \hat{B}\bar{X} \tag{3.5}$$

and

$$\hat{B} = \frac{\sum_{i=1}^{n} (X_i - \bar{X})(Y_i - \bar{Y})}{\sum_{i=1}^{n} (X_i - \bar{X})^2} \tag{3.6}$$

where the symbol "caret" ($\wedge$) denotes estimator and the symbol "overbar" $\bar{(\,)}$ denotes average value, and n denotes the total number of specimens.

The recommended expression for estimating the variance of the normal distribution for $\log(N)$ is:

$$\hat{\mu}^2 = \frac{\sum_{i=1}^{n} (Y_i - \hat{Y}_i)^2}{n - 2} \tag{3.7}$$

in which

$$\hat{Y}_i = \hat{A} + \hat{B}X_i \tag{3.8}$$

and the $(n - 2)$ term in the denominator is used instead of n to make $\hat{\mu}^2$ an unbiased estimator of the normal population variance μ^2.

An exact confidence band for the entire median S–N or ε–N curve (which means that all points on the linear or linearized S–N or ε–N curve are considered simultaneously) may be calculated using the following equation:

$$Y = \hat{A} + \hat{B}X \pm \sqrt{2F_p}\,\hat{\mu}\left[\frac{1}{n} + \frac{(X - \bar{X})^2}{\sum_{i=1}^{n} (X_i - \bar{X})^2}\right]^{1/2} \tag{3.9}$$

in which F_p is given in Tables [1] for reliability levels of 95% and 99%, although use of the latter is not recommended.

A process for verification of the linear material behavior hypothesis is also provided by the ASTM practice. Assuming that fatigue tests were conducted at l different stress levels, and that m_i specimen replications were observed per stress level, then the hypothesis of linearity is accepted if the following conditions holds true:

$$\frac{\sum_{i=1}^{l} m_i(\bar{Y}_i - \bar{\bar{Y}}_i)^2 \big/ (l-2)}{\sum_{i=1}^{l} \sum_{j=1}^{m_i} (\bar{Y}_{ij} - \bar{Y}_i)^2 \big/ (n-l)} \leq F_p \tag{3.10}$$

3.2.3 Whitney's Pooling Scheme

Whitney's method consists of a procedure allowing the generation of an S–N curve with some statistical value without resorting to an extremely large database. This method uses the "wear-out" or "strength degradation" model approach, initially proposed by Hahn and Kim [6], but follows an alternative mathematical procedure for estimation of the model parameters.

The method is based on two assumptions: a classic power law representation of the S–N, like Eq. 3.2, and a two-parameter Weibull distribution of time to failure.

For every stress level, the probability of survival after N cycles is given by a two-parameter Weibull distribution:

$$P_S(N) = \exp\left[-\left(\frac{N}{\bar{N}}\right)^{\alpha_f}\right] \tag{3.11}$$

with $\bar{N}$ and α_f, being the scale and shape parameters of the Weibull distribution.

The process for the derivation of the S–N curve comprises the following steps: a two-parameter Weibull distribution is fitted to the data of the ith stress level:

$$P_S(N_i) = \exp\left[-\left(\frac{N_i}{\bar{N}_i}\right)^{\alpha_{fi}}\right], \quad i = 1,\ldots,l \tag{3.12}$$

The parameters α_{fi} and $\bar{N}_i$ of each Weibull distribution are determined by solving the following set of equations for the Maximum Likelihood Estimators $\hat{\alpha}_{fi}$ and $\hat{\bar{N}}_i$:

$$\frac{\sum_{j=1}^{m_i} N_{ij}^{\hat{\alpha}_{fi}} \ln(N_{ij})}{\sum_{j=1}^{m_i} N_{ij}^{\alpha_{fi}}} - \frac{1}{m_i}\sum_{j=1}^{m_i} \ln(N_{ij}) - \frac{1}{\alpha_{fi}} = 0 \tag{3.13}$$

$$\hat{\bar{N}}_i = \left(\frac{1}{m_i}\sum_{j=1}^{m_i} N_{ij}^{\hat{\alpha}_{fi}}\right)^{1/\hat{\alpha}_{fi}} \tag{3.14}$$

with m_i being the number of specimens at each stress level. Equation (3.13) has only one positive root, which is the estimated value for the shape parameter α_{fi}. The resulting value of $\hat{a}_{fi}$ can be used in conjunction with Eq. 3.14 to obtain $\hat{\bar{N}}_i$.

Assuming that α_f is independent of the stress level, a data pooling technique can be used in order to determine a single value of α_f for the whole range of applied stresses [3]. According to [3], the pooled dataset is created by normalization of the number of cycles to failure for every specimen at each stress level with respect to the characteristic life of each stress level, i.e. the estimated scale parameter, $\hat{\bar{N}}_i$. The following normalized dataset is therefore formed:

$$Q(Q_{i1}, Q_{i2}, \ldots, Q_{im_i}), \quad i = 1, 2, \ldots, l \tag{3.15}$$

where:

$$Q_{ij} = \frac{N_{ij}}{\hat{\bar{N}}_i} \tag{3.16}$$

It is assumed that this set of data also follows a two-parameter Weibull distribution:

$$P_S(Q) = \exp\left[-\left(\frac{Q}{Q_o}\right)^{\alpha_f}\right] \tag{3.17}$$

with the values of the maximum likelihood estimators for the parameters, α_f and Q_o, estimated by solving the following set of equations:

$$\frac{\sum_{i=1}^{l}\sum_{j=1}^{m_i} Q_{ij}^{\hat{\alpha}_f}\ln(Q_{ij})}{\sum_{i=1}^{l}\sum_{j=1}^{m_i} Q_{ij}^{\hat{\alpha}_f}} - \frac{1}{n}\sum_{i=1}^{l}\sum_{j=1}^{m_i}\ln(Q_{ij}) - \frac{1}{\hat{\alpha}_f} = 0 \tag{3.18}$$

$$\hat{Q}_o = \left(\frac{1}{n}\sum_{i=1}^{l}\sum_{j=1}^{m_i} Q_{ij}^{\hat{\alpha}_f}\right)^{1/\hat{\alpha}_f} \tag{3.19}$$

As stated in [2, 3], the value of $\hat{Q}_o$ has to be unity for a perfect fit. If $\hat{Q}_o$ takes any value other than unity, the characteristic number of cycles for each stress level can be adjusted as:

$$\bar{N}_{oi} = \hat{Q}_o\hat{\bar{N}}_i \tag{3.20}$$

Substitution of $\hat{\bar{N}}_i$ in Eq. 3.16 by $\bar{N}_{oi}$ and repetition of the process leads to the estimation of $\hat{\alpha}_f$.

The slope of the S–N curve, Eq. 3.10, $1/k$ and the y-intercept, σ_o can be determined by fitting $\log(\sigma)$ versus $\log(\bar{N}_{oi})$ to a straight line. With σ_o, $1/k$ and α_f already determined, the S–N curve at any specified level of reliability can be calculated by:

$$\sigma_a = \sigma_o\left\{[-\ln(P_S(N))]^{\left(\frac{1}{\alpha_f k}\right)}\right\}N^{\left(-\frac{1}{k}\right)} \tag{3.21}$$

Censoring techniques are used for experimental data from specimens that have not failed after a predetermined number of cycles (run-outs). In this case, the equations for derivation of the maximum likelihood estimators for the parameters α_{fi} and $\bar{N}_i$ of the Weibull distributions at each stress level area as follows [2, 3]:

$$\frac{\sum_{j=1}^{r_i} N_{ij}^{\hat{\alpha}_{fi}} \ln(N_{ij}) + (m_i - r_i) N_{si}^{\hat{\alpha}_{fi}} \ln(N_{si})}{\sum_{j=1}^{r_i} N_{ij}^{\hat{\alpha}_{fi}} + (m_i - r_i) N_{si}^{\hat{\alpha}_{fi}}} - \frac{1}{r_i} \sum_{j=1}^{r_i} \ln(N_{ij}) - \frac{1}{\hat{\alpha}_{fi}} = 0 \qquad (3.22)$$

and

$$\hat{\bar{N}}_i = \left\{ \frac{1}{r_i} \left[\sum_{j=1}^{r_i} N_{ij}^{\hat{\alpha}_{fi}} + (m_i - r_i) N_{si}^{\hat{\alpha}_{fi}} \right] \right\}^{\frac{1}{\hat{\alpha}_{fi}}} \qquad (3.23)$$

In the above equations, m_i is the total number of specimens tested under the ith stress level, r_i is the number of failed specimens under that stress level and N_{si} is the number of cycles after which the test was stopped.

The parameters of the Weibull distribution of the pooled data are calculated by:

$$\frac{\sum_{i=1}^{l} \sum_{j=1}^{r_i} X_{ij}^{\hat{\alpha}_f} \ln(Q_{ij}) + \sum_{i=1}^{l} (m_i - r_i) Z_i^{\hat{\alpha}_f} \ln(Z_i)}{\sum_{i=1}^{l} \sum_{j=1}^{r_i} Q_{ij}^{\hat{\alpha}_f} + \sum_{i=1}^{l} (m_i - r_i) Z_i^{\hat{\alpha}_f}} - \frac{1}{r_T} \sum_{i=1}^{l} \sum_{j=1}^{r_i} \ln(Q_{ij}) - \frac{1}{\hat{\alpha}_f} = 0$$

$$(3.24)$$

$$\hat{Q}_o = \left\{ \frac{1}{r_T} \left[\sum_{i=1}^{l} \sum_{j=1}^{r_i} Q_{ij}^{\hat{\alpha}_f} + \sum_{i=1}^{l} (m_i - r_i) Z_i^{\hat{\alpha}_f} \right] \right\}^{\frac{1}{\hat{\alpha}_f}} \qquad (3.25)$$

The following notation was used:

$$Z_i = N_{si}/\hat{\bar{N}}_i$$

and $r_T = \sum_{i=1}^{l} r_i$ the total number of failed specimens.

3.2.4 Sendeckyj's Wear-Out Model

The method proposed by Sendeckyj [4] is based on the "wear-out" or "strength degradation" model according to which a direct relationship exists between static strength distribution, residual strength distribution of the material after being subjected to a specific load history and distribution of time to failure at a maximum stress level. According to the author, fatigue data can be placed in four major categories:

Static strength data: tensile or compressive static strength data. The loading rate should be comparable to the loading rate during fatigue testing in order to obtain comparable results.

Fatigue failure data: Data points obtained from specimens that failed after a determined number of fatigue cycles.

Run-outs: fatigue tests not continued up to failure are called "run-outs". They are characterized as fatigue failure data by the stress level and the corresponding number of cycles up to test termination. Data points in this class are either censored (not used at all) or tested under monotonic load to determine residual strength.

Residual strength data: residual static strength determined under a loading rate comparable to the cyclic loading rate. Every data point in this class corresponds to three values: residual strength, corresponding cyclic stress and number of loading cycles.

After considering the data from these four categories, the described fatigue model is based on three assumptions:

The S–N curve is described by a deterministic equation of any type, e.g., Lin–Log, Log–Log, etc.

There is a relationship between the static and fatigue strengths of the material in the sense that the stronger specimen in fatigue (the one with the longer lifetime for a given stress level) is assumed to also have the higher static strength. By using adequate fatigue model parameters, all fatigue data can be converted into equivalent static strength data.

The static strength data (including the converted fatigue data) follow a two-parameter Weibull distribution.

The wear-out model developed for the conversion of fatigue strengths into equivalent static strengths is mathematically described by the following deterministic equation:

$$\sigma_e = \sigma_{max}\left[(\sigma_r/\sigma_{max})^{\frac{1}{G}} + (N-1)C\right]^G \tag{3.26}$$

where σ_e is the equivalent static strength, σ_{max} denotes the maximum cyclic stress level, σ_r is the residual static strength (where applicable), N is the number of loading cycles and G and C are the fatigue model parameters to be determined.

In the case of fatigue failure, when $\sigma_r = \sigma_{max}$ and $N = N_f$ the above equation reduces to the following:

$$\sigma_e = \sigma_{max}(1 - C + N_f C)^G \tag{3.27}$$

To use the model, values for the G and C parameters must initially be assumed (or estimated) from the available fatigue data considering G as the slope of the S–N curve and C a constant that defines the shape of the S–N curve for the low-cycle fatigue region. Then, by using the selected G and C values, all data are converted into the equivalent static strengths, σ_e, using Eq. 3.26. A Weibull distribution is then fitted to the equivalent static strength dataset by using maximum likelihood estimators (MLE). The probability of survival of the Weibull distribution is given by:

$$P_S(\sigma_e) = \exp\left[-\left(\frac{\sigma_e}{\beta}\right)^{\alpha_f}\right]$$

(3.28)

This process is performed iteratively for different values of G and C until the maximum value of shape parameter a_f is obtained.

Using the selected/estimated set of parameters G, C and β, a_f the fatigue curve can be plotted for any desired reliability level $P_S(N)$ (including 50%, which denotes the mean value of experimental data) by using the following equation:

$$\sigma_{max} = \beta\left\{[-\ln(P_S(N))]^{\frac{1}{\alpha_f}}\right\}[(N - A)C]^{-G}$$

(3.29)

with

$$A = -\frac{1 - C}{C}$$

(3.30)

3.3 Statistical Analysis of Fatigue Data–Reliability-Based S–N Curves

The fatigue data presented in Chap. 2, which will be used in this paragraph for the demonstration of the applicability and evaluation of the modeling ability of the examined methods, were initially analyzed in order to validate that they conform to the replication guidelines for S–N testing given by ASTM and [7]. Replication is defined as:

$$\% \text{ replication} = 100\left[1 - \frac{\text{Total number of different stress levels used in testing}}{\text{Total number of specimens tested}}\right]$$

(3.31)

According to the ASTM E739 practice and [7], an appropriate number of specimens must be tested and a good replication should be achieved in order to assure that a random sample of the material is being tested. According to the aforementioned practice, numbers of specimens of between 12 and 24 are appropriate in order to obtain S–N curves with statistical significance. Replication of more than 75% is required for the derivation of reliability data, while a replication between 50 and 75% suffices for the derivation of datasets for design allowables. For the cases examined here, it was ensured that each dataset comprised numbers of specimens greater than 12, while replication was equal to 69.05 ± 3.57%.

Table 3.1 Fatigue model parameters

θ (deg)	R	NLD, Eq. 3.2 σ_o (MPa)	$1/k$	ASTM, Eq. 3.2 σ_o (MPa)	$1/k$	Whitney, Eq. 3.21 σ_o (MPa)	α_f	$1/k$	Sendeckyj, Eq. 3.29 β	α_f	G	C
0	0.5	388.48	0.05	393.29	0.05	363.11	1.17	0.04	235.84	33.23	0.20	1e-6
0	0.1	588.39	0.10	613.82	0.10	600.80	2.34	0.10	239.44	13.84	0.09	1e-4
0	−1	161.72	0.05	164.31	0.05	155.71	1.51	0.05	105.99	29.37	0.05	2e-4
0	10	288.67	0.05	304.52	0.05	265.11	1.42	0.04	289.61	28.47	0.05	1.50
15	0.1	186.84	0.08	195.35	0.08	181.73	2.12	0.07	190.85	28.46	0.07	1.29
30	−1	127.94	0.08	124.72	0.08	134.00	1.44	0.08	67.59	18.32	0.07	2e-4
30	10	337.13	0.11	355.31	0.12	347.08	4.41	0.11	382.19	32.09	0.12	1.46
45	0.5	178.61	0.07	182.02	0.07	172.79	2.07	0.07	101.72	32.65	0.08	2e-4
45	0.1	184.81	0.10	178.24	0.09	183.58	1.52	0.09	86.20	13.02	0.09	2e-4
45	−1	132.87	0.07	122.25	0.07	137.96	1.07	0.07	68.23	13.42	0.06	5e-5
45	10	259.72	0.08	257.12	0.08	276.54	1.71	0.08	139.05	21.92	0.09	2e-4
60	−1	162.26	0.12	152.58	0.11	142.28	1.90	0.10	55.65	15.13	0.11	5e-5
60	10	154.14	0.09	155.57	0.08	156.80	2.79	0.08	85.20	34.53	0.09	4e-4
75	0.1	085.16	0.08	088.63	0.08	083.82	2.06	0.08	86.73	26.31	0.08	1.44
90	0.1	059.87	0.07	062.16	0.07	059.20	3.10	0.07	37.49	34.26	0.07	8e-4
90	−1	115.75	0.10	109.82	0.09	092.78	1.17	0.07	94.71	16.06	0.07	1.46
90	10	091.42	0.07	078.22	0.05	077.17	1.57	0.05	51.96	29.24	0.05	2e-4

The four examined methods were applied on the material datasets presented in Chap. 2, and the estimated model parameters are given in Table 3.1. Observation of the results shows that all methods estimate similar slopes for the derived S–N curves ($1/k$ for NLD and ASTM, $1/k$ for Whitney and G for Sendeckyj models), while the y-intercept of the S–N curve cannot be directly compared, since for the two last models it is also a function of the estimated shape parameters, see Eqs. 3.21 and 3.29. A slope ($1/k$ in Eq. 3.2) between 0.05 and 0.12 was calculated, with an average of 0.077 and a standard deviation of ±0.02 (according to the ASTM method) for all the studied combinations of material configuration and loading pattern. It is observed that for $R = 0.5$, independent of specimen configuration, the slope of the S–N curves is lowest when compared to the slopes of S–N curves for the other loading patterns. This indicates that the cyclic stress amplitude (which is minimum for $R = 0.5$) is a very critical parameter for the lifetime of the examined composite materials. These findings are in agreement with results of studies found in the literature for similar material systems and loading cases, e.g., [8–12]. The available static strength data were not considered for the application of the fatigue models and the estimation of their parameters. Therefore, the obtained S–N curves are valid, without the need for any extrapolation, for numbers of cycles between ca 1,000 and 2,000,000.

Selected S–N curves, typical of all datasets, covering on-and off-axis angles in the range between 0° and 90° and different applied fatigue loading patterns, are presented in Figs. 3.1, 3.2, 3.3, 3.4, 3.5, 3.6, 3.7, 3.8, 3.9 for demonstration.

Fig. 3.1 Reliability-based
S–N curves for on-axis
specimens, $R = 10$

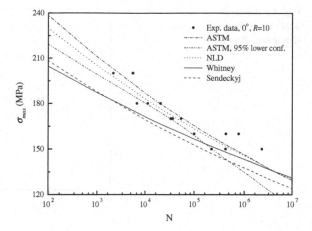

Fig. 3.2 Reliability-based
S–N curves for 15° off-axis
specimens, $R = 0.1$

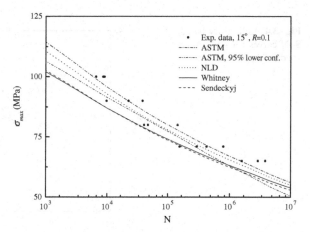

The median curve, according to ASTM, is indicated by a dash-dot line, accompanied by the 95% lower confidence limit (indicated by a dash-dot-dot line) according to the same practice, in all figures. The "Whitney" S–N curve for 95% reliability is indicated by a solid line, the S–N curves for 95% obtained by the method that assumes normal life distribution (NLD) are indicated by dots, while the S–N curves for 95% reliability level, derived according to the Sendeckyj wearout model, are shown by dashed lines.

Comparison of the estimated S–N curves as derived from the different fatigue models leads to the following conclusions:

The method based on the normal lifetime distribution and denoted by "NLD" in the graphs proved to be non-conservative, producing S–N curves situated above all the others on the S–N plane. In most of the examined cases it derived S–N curves that are better corroborated by the median S–N curve, indicated by ASTM in the graphs, than the 95% reliability one. Therefore, this simplified method cannot be applied for the examined material systems.

Fig. 3.3 Reliability-based
S–N curves for 30° off-axis
specimens, $R = 10$

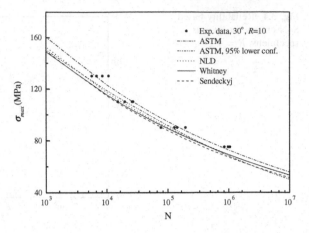

Fig. 3.4 Reliability-based
S–N curves for 45° off-axis
specimens, $R = 10$

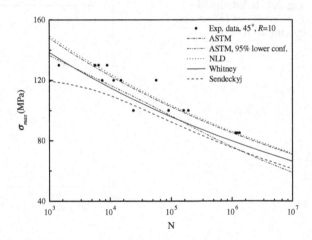

Fig. 3.5 Reliability-based
S–N curves for 60° off-axis
specimens, $R = 10$

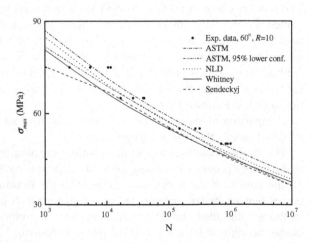

Fig. 3.6 Reliability-based
S–N curves for 75° off-axis
specimens, $R = 0.1$

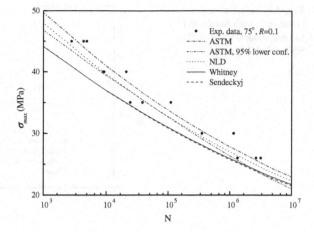

Fig. 3.7 Reliability-based
S–N curves for 90° off-axis
specimens, $R = 0.1$

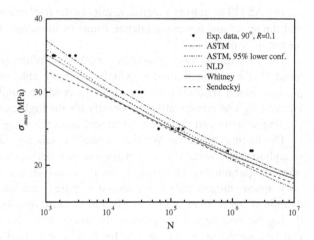

Fig. 3.8 Reliability-based
S–N curves for 90° off-axis
specimens, $R = -1$

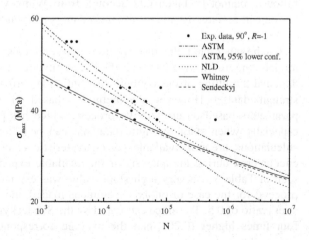

Fig. 3.9 Effect of using
static data with Sendeckyj
fatigue model, $0°$, $R = 0.5$,
95% reliability level

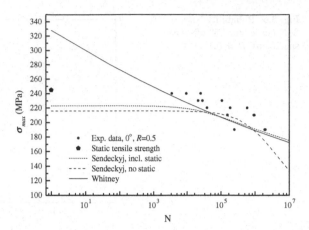

The ASTM directives practice results in the derivation of the median S–N curve and the upper and lower confidence limits inside which the median line should lie on the S–N plane.

The remaining two examined methods, Whitney's pooling scheme and Sendeckyj's wear-out model, which are both able to derive S–N curves for predetermined reliability levels, seem to produce similar S–N curves, with the latter being less conservative, especially for the high-cycle fatigue range. Both the pooling scheme and the wear-out model have advantages and disadvantages.

The method based on Whitney's pooling scheme has the advantage of being straightforward, since it does not require any optimization process for estimation of model parameters. However, it has the disadvantage that it cannot be applied if only sparse fatigue data exist and it requires multiple fatigue results for each examined stress level in order to fit the Weibull distribution to each one of them during the first step of the process. In this context, the static strength data, which can be considered as fatigue data for $N = 1$ or $¼$ (a discussion on this argument follows), cannot be taken into account from Whitney's method since no distribution can be fitted to this set of stress data corresponding to the same number of cycles.

On the other hand, this inconvenience does not exist in the case of the wear-out model, since initially all fatigue data are converted to equivalent static strength data and the fitting of the Weibull distribution is performed on this newly created strength dataset. However, the method is based on optimization of the model parameters based on an iterative process, which has proved rather troublesome, especially when static strength data that can be taken into account during the calculations are not available. A characteristic example of this problem was encountered during the analysis of the available experimental data. As it can be seen in Table 3.1, a very high slope value was estimated (after the optimization process) for the case of on-axis specimens loaded under tensile fatigue loads at a stress ratio of 0.5. The slope estimated by the Sendeckyj model was approximately four times higher (0.20) than the average corresponding value from all other models (0.05). As a result, a very steep curve, especially for fatigue that lives longer than 100,000, was derived, as presented in Fig. 3.9.

Normally, constant amplitude fatigue data are collected in the range between 100 and 10^7 loading cycles. Therefore, a regression is necessary for modeling of the fatigue life of the examined material in the very low-cycle fatigue region.

The use of the static strength data can, theoretically, be helpful for this regression, since they can be considered as the fatigue strength for one cycle until failure.[1]

Use of the static strength data and considering them as fatigue failures corresponding to one cycle can significantly improve the results for the examined case as shown in the same figure. The estimated parameters from the Sendeckyj model for this case were: $\beta = 242,78$, $\alpha_f = 34.57$, $G = 0.038$ and $C = 5.7e{-}5$. Use of the static strength data in the formulation derives an S–N curve that is much less conservative at the low- and high-cycle regimes for the examined material. As presented in Fig. 3.9, the S–N curve derived by using the static strength data is well corroborated by the Whitney curve for numbers of cycles higher than 10,000, while the curves diverge at the low-cycle regime.

However, it is still uncertain as to whether static data can be mixed with fatigue data for reasons such as the following:

• The static strength data are, in most cases, obtained at strain rates much lower than the maximum strain rates in fatigue loading.
• Even in cases when the static strength data were collected under strain rates comparable to the corresponding rates in fatigue loading, the assumption that the static strength data should be part of the S–N determination process has not been validated, e.g., [13]. An increase in static strength of around 70% was reported in Chap. 2 of this volume when the quasi-static tests were performed under a "fatigue strain rate". As mentioned in [13] an increase of around 35–40% has been observed for similar GFRP laminates. This was higher than that predicted by a Lin–Log regression line excluding static data, but considerably lower than what is suggested by extrapolating a Log–Log regression line excluding static data to $N = 1$.
• Although the use of static strength data, especially if data at a "fatigue strain rate" exist, can improve modeling in the low-cycle fatigue range, it affects the entire S–N curve and therefore the regression for the high-cycle fatigue range without any justification.
• The failure modes under quasi-static loading are in general different from the fatigue failure mode and therefore it is questionable whether the static strength data can be considered as valid fatigue data or not.

---

[1] It can even be argued that this should be 0.25 cycles to failure, or 0.75 cycles to failure, if the load is sinusoidal starting with tension, and the failure mode is tensile or compressive, respectively.

Therefore, the decision as to whether or not to use the static strength data should be justified after taking into consideration the differences between the static and the fatigue failure mechanisms exhibited by composite materials.

3.4 Conclusions

Four methods for the statistical analysis of composite material fatigue data and derivation of reliability-based S–N curves have been presented in this chapter. All are based on assumptions related to the distribution of the lifetime data (numbers of cycles or their logarithms) or the distribution of the fatigue strength and the type of the S–N curve that can best represent the fatigue behavior.

Two of the examined methods seem to be more adequate for the modeling of the fatigue behavior of the examined composite material system; the Whitney's pooling scheme and the wear-out model introduced by Sendeckyj. Although both have disadvantages, they offer a consistent, reliable and accurate algorithm for the derivation of S–N curves at any reliability level. Whitney's method is straightforward and does not require any optimization process for estimation of the model parameters. Sendeckyj's method is more complicated but has the merit that it can be used even when only sparse fatigue data exist.

There is as yet no evidence that the static strength data can be safely used together with fatigue strength data for the derivation of an S–N curve. However, when low-cycle fatigue data are limited or absent, and the low-cycle fatigue regime is important for the application, the elaboration of the static strength can assist with the estimation of the fatigue model parameters.

References

1. ASTM E739 – 91 Standard Practice for Statistical Analysis of Linear or Linearized Stress-Life (S–N) and Strain-Life (ε-N) Fatigue Data (2004)
2. J.M. Whitney, Fatigue characterization of composite materials. in *Fatigue of Fibrous Composite Materials, ASTM STP 723*, American Society for Testing and Materials (1981), pp. 133–151
3. J.M. Whitney, I.M. Daniel, R.B. Pipes, *Experimental Mechanics of Fiber Reinforced Composite Materials* (Prentice-Hall, Englewood Cliffs, 1984)
4. G.P. Sendeckyj, Fitting models to composite materials, in *Test Methods and Design Allowables for Fibrous Composites, ASTM STP 734*, ed. by C.C. Chamis (American Society for Testing and Materials, West Conshohocken, 1981), pp. 245–260
5. L. Young, J.C. Ekvall, Reliability of fatigue testing, in *Statistical Analysis of Fatigue Data, ASTM STP 744*, ed. by R.E. Little, J.C. Ekvall (American Society for Testing and Materials, West Conshohocken, 1981), pp. 55–74
6. H.T. Hahn, R.Y. Kim, Fatigue behavior of composite laminate. J. Compos. Mater. **10**(2), 156–180 (1976)
7. R.E. Little, *Manual on Statistical Planning and Analysis. STP 588* (American Society for Testing and Materials, West Conshohocken, 1975), pp. 46–60

8. P.W. Bach, Glass and Hybrid Fibre Performance. in *Design of Composite Structures Against Fatigue.* ed by R.M. Mayer. Applications to Wind Turbine Blades, (1996)
9. B. Harris, N. Gathercole, H. Reiter, T. Adam, Fatigue of carbon-fibre-reinforced plastics under block-loading conditions. Compos. Part A Appl. S **28A**, 327–337 (1997)
10. R.M. Mayer, Benchmark Tests in *Design of Composite Structures Against Fatigue.* ed. by R.M. Mayer. Applications to Wind Turbine Blades, (1996)
11. J.F. Mandell, D.D. Samborsky, DOE/MSU composite material fatigue database: Test Methods Material and Analysis, Sandia National Laboratories/Montana State University, SAND97–3002, (online via www.sandia.gov/wind, last update, v. 18.1, 25 March 2009)
12. R.P.L. Nijssen, OptiDAT–fatigue of wind turbine materials database (2006) http://www.kc-wmc.nl/optimat_blades/index.htm
13. R.P.L. Nijssen, Tensile tests on standard OB specimens–effect of strain rate, OB_TG1_R014:2004. http://www.wmc.eu/public_docs/10221_003.pdf

Chapter 4
Modeling the Fatigue Behavior of Fiber-Reinforced Composite Materials Under Constant Amplitude Loading

4.1 Introduction

The fatigue design of a structural component is based on the evaluation of the fatigue behavior of the constituent materials under loading conditions similar to those of the structural component that it will encounter during its operational life. The failure mechanisms of a fiber-reinforced composite material are more complex than those of a metallic one. A synergy of matrix cracks, layer delamination and finally fiber failure comprises the basic failure mechanisms of fiber-reinforced composite materials. Due to this complexity, the application of a simple theory for the quantification of the effect of each one of these failure mechanisms on the fatigue life of a composite material and the eventual establishment of a method for the derivation of design allowables is very difficult. On the other hand, the measurement of a number of macroscopic damage metrics that are affected by the development of the failure mechanisms is possible and requires only simple experimental procedures and set-ups. For example, the relationship between the applied load and the corresponding number of cycles up to failure, the remaining strength after a fatigue loading or the stiffness degradation during the application of a constant or variable amplitude loading pattern can be used for the derivation of reliable allowable values for the design of a structural component.

Significant research efforts have been devoted to the understanding of the fatigue behavior of fiber-reinforced composite materials and development of techniques to model the fatigue life and predict material behavior under different conditions. A large amount of fatigue data have been derived for composite materials under different loading patterns in several engineering domains, such as the aerospace and wind turbine industries, e.g., [1–5] during the last decades. Usually, the results are presented in terms of S–N curves (of any type), and the design allowables are defined based on the life of the examined material under specific applied loads. Depending on the nature and amount of available data, design allowables in terms of S–N curves can also be determined for any reliability

A. P. Vassilopoulos and T. Keller, *Fatigue of Fiber-reinforced Composites*,
Engineering Materials and Processes, DOI: 10.1007/978-1-84996-181-3_4,
© Springer-Verlag London Limited 2011

level, taking into account the stochastic nature of the phenomenon, as described in Chap. 3 of this volume. The main drawback of this type of fatigue data interpretation is the necessity for the failure of the material in order to derive the S–N curve. This is based on destructive testing and there is no way to implement this method during fatigue loading to serve as a life-monitoring tool. A different damage metric must be used for this purpose. The behavior of this damage metric during fatigue life must be characterized with simple and limited experiments, without the need to reach material failure, so that the method can be used for estimation of the fatigue life of the examined material or structural component while it is in service.

A damage metric that fulfills these requirements is the stiffness of the material. Use of stiffness as a damage metric does not require the failure of the material since it can be measured in a non-destructive way. Moreover, as the stiffness of composite materials exhibits less scatter than strength, the modeling of the stiffness behavior during fatigue life can be performed based on fewer experimental data than those needed for a reliable statistical analysis of the fatigue behavior of the examined material based on the derived S–N curves. As it will be presented later on, the stiffness degradation measurements can be used for the derivation of S–N curves that do not correspond to failure, but to a stiffness degradation level. This kind of curve is very useful for the design of structures that comprises of moving parts, since it can constitute design allowables that conforms with specified limits of displacement (strains) in order to maintain the structure's geometry and avoid causing problems in operation either by conflicting movements between different parts or by undesirable changes in the aerodynamics of each structural component.

Regardless of the method used for modeling, the established model in terms of a mathematical expression can be used to interpolate or extrapolate the fatigue behavior for different numbers of fatigue cycles. However, the situation is not as simple for the modeling of the behavior of a composite material when it is loaded under different loading patterns, e.g., Tension-Tension (T-T), tension-compression (T-C) or Compression-Compression (C-C) fatigue. The effect of the different mean stress levels of the various loading cases is very critical for the fatigue life of any composite material. It is not easy to interpolate between different loading domains in order to model the behavior of the material under new loadings and the constant life diagrams (CLDs) were established to address this problem. Depending on the CLD formulation, the interpolation between known material behaviors can be linear or non-linear. However, the accuracy of a CLD can only be evaluated based on comparisons with available fatigue data for the examined material.

In this chapter, the types of S–N curves commonly used for composite materials are reviewed and a critical comparison of their modeling ability is made. The models based on the linear regression of the stress vs. the logarithm of the number of cycles to failure (Lin-Log) or the logarithm of the applied cyclic stress vs. the logarithms of the number of cycles to failure (Log–Log) is presented and their modeling accuracy is compared to the modeling accuracy of novel methods for the interpretation of fatigue data based on stochastic computational tools, such as artificial neural networks and genetic programming, as presented in [6–8].

The most commonly used constant life diagrams for composite materials and those most recently introduced are also presented in this chapter. Their predictive ability is evaluated by using the dataset in Chap. 2 and others from the literature. The applicability of the models, the need for experimental data and the accuracy of the predictions are considered as critical parameters for the evaluation.

The stiffness degradation during the fatigue life of the examined composite material was thoroughly examined in Chap. 2 as a function of several parameters: the loading case (C-C, T-T or T-C loading), the applied cyclic stress level and the percentage of fibers along the loading direction. A systematic statistical analysis, corresponding to a certain R-value and off-axis direction, proved that irrespective of the constant amplitude stress level, stiffness degradation data are satisfactorily fitted by standard statistical distributions. Modeling of the stiffness degradation can be used for the derivation of S–N curves that correspond to specific stiffness degradation levels and not to failure. Here, these stiffness-controlled curves, designated Sc-N curves, were determined for each R-ratio and off-axis direction and compared to fatigue strength curves. It is shown that these two kinds of curves can be correlated and it is thus possible to derive fatigue design allowables corresponding to specific levels of stiffness degradation and survival probability. Furthermore, even by using only half of the experimental data, Sc-N curves can still be accurately defined.

4.2 Which Type of S–N Curve?

One of the most explicit and straightforward ways to represent experimental fatigue data is the S–N diagram. It is preferred to other approaches for the modeling of the fatigue life of FRP composite materials, e.g., those based on stiffness degradation, or crack propagation measurements during lifetime, since it requires input data (applied load and corresponding cycles to failure) that can be collected using very simple recording devices.

Usually, fatigue data for preliminary design purposes are gathered in the region of fatigue cycles ranging between 10^3 and 10^7. However, depending on the application, high- or low-cycle fatigue regimes can be of interest. Additional data are needed in such cases to avoid the danger of poor modeling due to extrapolation in an unknown space. Although for the high-cycle fatigue regime long-term fatigue data must be acquired, the situation seems easier for low-cycle fatigue where static strength data can apparently be used in combination with the pure fatigue data. However, when the static strength data are considered in the analysis, other problems arise, and thus the use of quasi-static strength data for the derivation of fatigue curves (such as fatigue data for 1 or ¼ cycle) is arguable. No complete study on this subject exists. Previous publications, e.g., [9], showed that quasi-static data should not be a part of the S–N curve, especially when they have been acquired under strain rates much lower that those used in fatigue loading. The use of quasi-static data in the regression leads to incorrect slopes of the S–N curves as presented in [9]. On the other hand, although excluding quasi-static data improved the description of the fatigue data, it introduced errors in the lifetime predictions when the low-cycle regime is important, as for example for loading spectra with a few high-load cycles.

The selection of the fatigue model that is established by fitting a mathematical equation to the experimental data is of paramount importance for any fatigue analysis. The fatigue model reflects the behavior of the experimental data to theoretical equations which are consequently used during design calculations. A number of different types of fatigue models (or types of S–N curves) have been presented in the literature, with the most "famous" being the empirical semi-logarithmic and logarithmic relationships. Based on these it is assumed that the logarithm of the loading cycles is linearly dependent on the cyclic stress parameter, or its logarithm. Fatigue models determined in this way do not take different stress ratios or frequencies into account, i.e., different model parameters should be determined for different loading conditions. Also, they do not take into account any of the failure mechanisms that develop during the failure process and these fatigue models therefore have the disadvantage of being case-sensitive, since they may derive very accurate modeling results for one material system under specified loading conditions but very poor results for another. Other more sophisticated fatigue formulations that also take the influence of stress ratio and/or frequency into account were also reported [10, 11]. A unified fatigue function that permits the representation of fatigue data under different loading conditions (different R-ratios) in a single two-parameter fatigue curve was proposed by Adam et al. [10]. In another work by Epaarachchi et al. [11], an empirical model that takes into account the influence of the stress ratio and loading frequency was presented and validated against experimental data from different glass fiber-reinforced plastic composites. Although these models seem promising, their empirical nature is a disadvantage as their predictive ability is strongly affected by the selection of a number of parameters that must be estimated or even, in some cases, assumed.

Methods for the S–N curve modeling of composite materials, also appropriate for the derivation of S–N curves that take into account the stochastic nature of the examined materials, have been established to permit the derivation of S–N curves with some statistical significance based on limited datasets e.g., [12, 13]. These statistical methods (already presented in detail in Chap. 3) are also based on a deterministic S–N equation for representation of the fatigue data; however a more complicated process, compared to the simple regression analysis, is followed for the estimation of model parameters. In addition, the methods introduce assumptions that also allow the run-outs (data from specimens that did not fail during loading) to be considered in the analysis.

Recently, artificial intelligence methods have been adopted for interpretation of the fatigue data of composite materials. Artificial intelligence methods have previously been used and validated in various fields. They appear to offer a means of dealing with many multivariate problems for which no exact analytical model exists or is very difficult to develop. Artificial neural networks (ANNs) have proved as very powerful tools for pattern recognition, data clustering, signal processing etc. During the last 10 years, ANNs have been used to model the fatigue life of composite materials by modeling their S–N behavior, e.g. [14–17], or by deriving constant life diagrams in order to model the effect of different stress ratios on the fatigue life of an FRP composite system, see e.g., [18, 19]. They have also been used in other engineering problems, such as prediction of the multiaxial strength of composite materials, Lee et al. [20], or the modeling of the fatigue crack growth rate of bonded FRP-wood interfaces, Jia et al. [21]. A hybrid neuro-fuzzy method, designated ANFIS (Adaptive Neuro-Fuzzy Inference System), has been used to model the fatigue life of unidirectional and multidirectional composite laminates. Results of its application to two, different in general, material systems have been presented in [6, 22]. Finally, Genetic Programming (GP) has been successfully used as a tool for the derivation of S–N curves and modeling of the fatigue behavior of composite materials, as presented by Vassilopoulos et al. [8].

This innovative tool in the field of fatigue life modeling, based on genetic programming, is also introduced in this chapter and its ability to derive appropriate S–N curves is evaluated against traditional empirical and statistical methods. Selected experimental data from the database presented in Chap. 2 have been used for the application of the different models and comparison of their modeling ability. For the application of the GP method the GP software tool from RML Technologies, Inc., Discipulus^{TM} [23] was used.

4.2.1 Empirical and Statistical S–N Formulations

As mentioned in Chap. 3, the easiest way to estimate the parameters of a fitted line representing material behavior is linear regression analysis, which can be performed even based on hand calculations. The resulting S–N curve yields an estimate of the mean time to failure as a function of the corresponding stress variable. In the S–N formulation, the stress variable σ can refer to any cyclic stress definition, σ_{max} (maximum cyclic stress), σ_a (cyclic stress amplitude) or even $\Delta\sigma$ (cyclic stress range). The mathematical expressions of the semi-logarithmic and logarithmic S–N formulations are given in Eqs. 4.1, 4.2:

$$\log(N) = A + B\sigma \ or \tag{4.1}$$

$$\sigma = \sigma_o N^{\left(-\frac{1}{k}\right)} \tag{4.2}$$

in which σ represents the selected stress variable and N the corresponding number of cycles to failure, while A, B, σ_0 and k are the model parameters that will be determined after the application of the process to the available fatigue data. The fatigue life is considered as the dependent variable whereas the stress or strain is considered as the independent variable for the fitting.

The appropriate formulations, Eq. 4.3, for Whitney's Weibull statistics, and Eq. 4.4 for the Sendeckyj's wear-out model were also used in this chapter for the derivation of S–N curves, corresponding to a 50% reliability level (see Chap. 3 for definition of the model parameters).

$$\sigma_a = \sigma_o \left\{ [-\ln P_S(N)]^{\left(\frac{1}{\gamma k}\right)} \right\} (N)^{\left(-\frac{1}{k}\right)} \tag{4.3}$$

$$\sigma_{max} = \beta \left[-\ln P_S(N)^{\frac{1}{\gamma}} \right] [(N-A)C]^{-G} \tag{4.4}$$

4.2.2 *Genetic Programming for Fatigue Life Modeling*

Genetic programming (GP) is a domain-independent problem-solving technique in which computer programs are evolved to solve, or approximately solve, problems. Genetic programming is a member of the broad family of techniques known as Evolutionary Algorithms. All these techniques are based on the Darwinian principle of reproduction and survival of the fittest and are similar to biological genetic operations such as crossover and mutation. Genetic programming addresses one of the central goals of computer science, namely automatic programming, which is to create, in an automated way, a computer program that enables a computer to solve a problem [24].

In genetic programming, the evolution operates on a population of computer programs of varying shapes and sizes. These programs are habitually represented as trees, as for example the one shown in Fig. 4.1, where the function $f(x) = 3r + ((x+5) - 2y)$ is represented in tree format. The operations are performed in the "tree branches" and the result is given at the "tree root".

Genetic programming starts with an initial population of thousands or millions of randomly generated computer programs composed of random choices of the primitive functions and terminals as defined in the first and second preparatory steps (see below) and then applies the principles of biological evolution to create a new (and often improved) population of programs. This new population is

Fig. 4.1 Tree representation in genetic programming

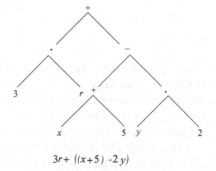

$$3r + ((x+5) - 2y)$$

generated in a domain-independent way using the Darwinian principle of survival of the fittest, an analogue of the naturally occurring genetic operation of crossover (sexual recombination), and occasional mutation [25]. The crossover operation is designed to create syntactically valid offspring programs (given closure amongst the set of programmatic ingredients). Genetic programming combines the expressive high-level symbolic representations of computer programs with the near-optimal efficiency of learning of Holland's genetic algorithm. A computer program that solves (or approximately solves) a given problem often emerges from this process [25].

Six major preparatory steps must be performed before applying genetic programming [25] in a given problem. These steps include preparation of datasets, setting-up of the model and design of the termination criteria, as explained in the following:

1. Determination of the *set of terminals*. The terminals can be seen as the inputs to the as-yet-undiscovered computer program. The set of terminals (or *Terminal Set T*, as it is often called) together with the set of functions is the ingredients from which genetic programming constructs a computer program to solve, or approximately solve, the problem.

2. Determination of the *set of primitive functions*. These functions will be used to generate the mathematical expression that attempts to fit the given finite sample of data. Each computer program is a combination of functions from the function set F and terminals from the terminal set T. The selected function and terminal sets should have the closure property in order that any possible combination of functions and terminals produces a valid executable computer program (a valid model).

3. Determination of the *fitness measure* which drives the evolutionary process. Each individual computer program in the population is executed and then evaluated, using the fitness measure, to determine how well it performs in the particular problem environment. The nature of the fitness measure varies with the problem: e.g., for many problems, fitness is naturally measured by the discrepancy between the result produced by an individual candidate program and the desired result; the closer this error is to zero, the better the program. For some problems, it may be appropriate to use a multi-objective fitness measure incorporating a combination of factors such as correctness, parsimony (smallness of the evolved program), and efficiency.

4. Determination of the *parameters for controlling the run*. These parameters define the guidelines in accordance with which each GP model that evolves. The population size, i.e., the number of created computer programs, the maximum number of runs, i.e., evolved program generations and the values of the various genetic operators are included in the list of parameters.

5. Determination of *the method for designating a result*. A frequently used method for result designation of a run is to appoint the best individual obtained in any generation of the population during the run, (i.e., the *best-so-far individual*) as being the result of the run.

6. Determination of *the criterion for terminating a run*. The maximum number of generations, or the maximum number of successive generations for which no improvement is achieved (values that were determined in step 4), is usually considered as the termination criteria.

Typically, the size of each developed program is limited, for practical reasons, to a certain maximum number of points (i.e., total number of functions and terminals) or a maximum depth of the program tree. Each computer program in the population is executed for a number of different *fitness cases* so that its fitness is measured as a sum or an average over a variety of different representative situations. For example, the fitness of an individual computer program in the population may be measured in terms of the sum of the absolute value of the differences between the output produced by the program and the correct answer (desired output) to the problem (i.e., the Minkowski distance) or the square root of the sum of the squares (i.e., Euclidean distance). These sums are selected from a sampling of different inputs (fitness cases) to the program. The fitness cases may be chosen at random or in a structured way (e.g., at regular intervals) [25].

All the individual programs of the initial population (generation 0) usually have exceedingly poor fitness, although some individuals in the population will fit the input data better than others. These differences in performance are then exploited by genetic programming. The Darwinian principle of reproduction and survival of the fittest and the genetic operations of crossover and mutation are used to create a new offspring population of individual computer programs from the current population.

The reproduction operation involves selecting a computer program from the current population of programs based on fitness (i.e., the better the fitness, the more likely the individual is to be selected) and allowing it to survive by copying it into the new population. Therefore, a new population, that of the offspring programs, replaces the old population. This iterative process is continued (new generations are evolved based on crossover and mutation operations) until a termination criterion, as defined in the sixth preparatory step, is satisfied. Regardless of their fitness, all generated programs, including those of the initial population, are syntactically valid executable programs.

The crossover operation creates new offspring computer programs from two parent programs selected on the basis of their fitness. The parent programs in genetic programming are usually of different shapes and sizes. The offspring programs are composed of sub-expressions from their parents. These offspring programs are usually of different shapes and sizes than their parents. For example, consider the two parent computer programs (models) represented as trees in Fig. 4.2 One crossover point is randomly and independently chosen in each parent. Consider that these crossover points are the division operator (/) in the first parent (the left one) and the multiplication operator (•) in the second parent (the right one). These two crossover fragments correspond to the underlying sub-programs (sub-trees) in the two parents–the sub-trees circled in Fig. 4.2. The two offspring resulting from the crossover operation, depicted in Fig. 4.3, have been created by

Fig. 4.2 Crossover operation: the two parent programs in tree representation

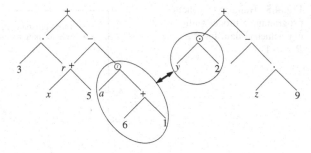

Fig. 4.3 Crossover operation: the two offspring programs in tree representation

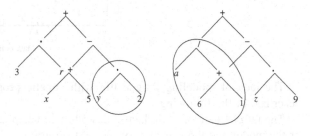

Fig. 4.4 Mutation operation of a parent program

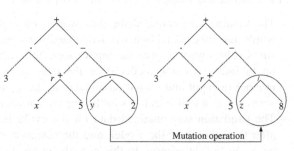

swapping the two sub-trees between the two parents in Fig. 4.2. Thus, the crossover operation creates new computer programs using parts of existing parent programs. Since entire sub-trees are swapped, the crossover operation always produces syntactically and semantically valid programs as offspring, regardless of the choice of the two crossover points. Because programs are selected to participate in the crossover operation based on their fitness, crossover allocates future trials to regions of the search space whose programs contain parts from promising programs [25]. The crossover is the predominant operation in GP.

The mutation operation creates an offspring computer program from one parent program that is selected based on its fitness. One sub-tree is randomly and independently chosen and then substituted with another sub-tree, see Fig. 4.4, by using the same growth procedure as it was originally used to create the initial random population. There are several kinds of mutations possible. Some examples are [26] the branch-mutation, where a complete sub-tree is replaced by another, the node-mutation, which applies a random change to a single node, etc. This asexual mutation operation is typically performed sparingly during each generation.

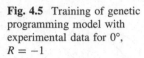

Fig. 4.5 Training of genetic programming model with experimental data for 0°, $R = -1$

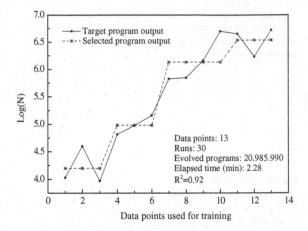

Data points: 13
Runs: 30
Evolved programs: 20.985.990
Elapsed time (min): 2.28
$R^2 = 0.92$

Data points used for training

The idea of modeling fatigue life with genetic programming was applied as described in the following:

The fatigue data of each dataset described in Chap. 2 was used for the training of the model and the selection of the best program.

- The training set contained the data which the tool used for learning. In other words, the fitness function was calculated based on the training set. Maximum stress values were used as the input parameters, while the corresponding cycles up to failure were considered as the desired output parameters. Given the number of input and output parameters in the training set, the process is characterized as a non-linear stochastic regression analysis.
- The validation set contained data for the evaluation of the evolved programs, after the training of the model, and the selection of the best one among them, based on the set criteria. In this case the criterion was the minimization of the error between the targeted output and the output produced by the evolved program. It is imperative that the validation set should contain examples that comprises of a good representative set of samples from the training domain.
- A test, or applied set, is constructed in the sequel, containing input data for which the output is expected to be calculated by the selected evolved program. Although these data are "new" and have not been used for the training or validation of the model, they should be in the range of the training set, since the ability of GP for extrapolation outside the training set has not yet been validated. For the case studied here, the test datasets were prepared so as to cover all the range from minimum to maximum cyclic stress levels in order to obtain, after the termination of each run, an entire S–N curve.

The same model (the selected evolved program) can be stored and potentially used to predict other output values for a new applied input dataset.

The training efficiency of the genetic programming tool was very good. As shown in Figs. 4.5, 4.6, 4.7, 4.8 where the target output is compared with the best program output after the training process for selected cases, the coefficient of

Fig. 4.6 Training of genetic programming model with experimental data for 15°, $R = 0.1$

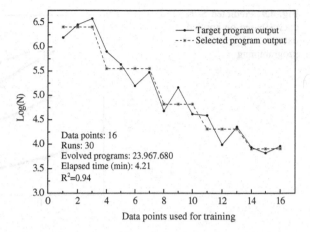

Fig. 4.7 Training of genetic programming model with experimental data for 45°, $R = 10$

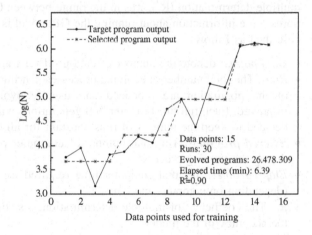

Fig. 4.8 Training of genetic programming model with experimental data for 75°, $R = 0.1$

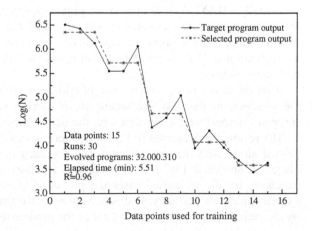

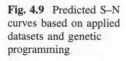

Fig. 4.9 Predicted S–N curves based on applied datasets and genetic programming

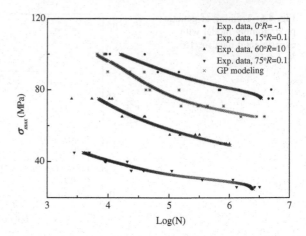

multiple determination (R^2), was in the range between 0.90 and 0.96. In the same Sures some information about running the GP model is also presented and can be described as follows:

- *Data points*: denote the number of data used for training the GP tool.
- *Runs*: The total number of evolving processes. During each run, new generations of the population are produced and usually evolve, i.e., the accuracy is improved. Each run finishes after 300 generations without improvement. It was decided to keep the number of runs constant for all examined cases.
- *Evolved programs*: The total number of computer programs that was evolved during the training process.
- *Elapsed time*: The total computer time required for the training of the model (depending on computer model).
- R^2: The coefficient of multiple determination, a statistical indicator that shows the accuracy of the model.

The GP tool was executed on an INTEL® Core™ i5 CPU 750 at 2.67 GHz, with 4 GB of RAM. As it can be seen in Figs. 4.5, 4.6, 4.7, 4.8 the data points used for each material case were preconditioned in different ways, sorted in ascending or descending order to prove the insensitivity of the proposed modeling technique to the treatment of the input data and also to avoid the generation of supervised modeling results.

After the development of a number of computer programs (during training) and the selection of the best one among them (during validation), the predictions (program output) were compared with the actual experimental data.

The results are presented in Fig. 4.9, where selected predicted S–N curves are plotted along with the experimental data for each material system on the S–N plane. As shown in Fig. 4.9, the modeling accuracy of genetic programming is excellent. In all the studied cases the produced curves follow the trend of the experimental data perfectly. It should be mentioned that the S–N curve predicted by the genetic programming tool is not of the predetermined type: power curve, or

Fig. 4.10 Third and fourth order polynomial S–N curve equations for GP output

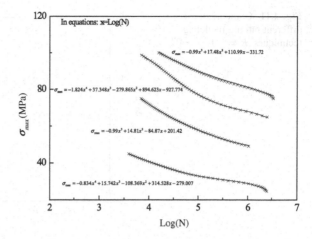

Table 4.1 Calculated fatigue model parameters for selected cases from the database presented in Chap. 2

	Linear regression, Eq. 4.2		Whitney, Eq. 4.3			Sendeckyj, Eq. 4.4			
Material	σ_o (MPa)	$1/k$	α_f	σ_o (MPa)	$1/k$	α_f	β (MPa)	G	C
$0°, R = -1$	164.31	−0.0517	1.51	155.71	−0.0463	29.37	105.10	−0.0476	1.54e−4
$15°, R = 0.1$	195.35	−0.0774	2.12	181.73	−0.0694	28.46	190.85	−0.0721	1.29
$60°, R = 10$	155.77	−0.0842	2.79	156.80	−0.0833	34.53	85.20	−0.0931	3.85e−4
$75°, R = 0.1$	88.63	−0.0838	2.06	83.82	−0.0769	26.31	86.73	−0.0781	1.44

polynomial, or semi-logarithmic etc. The resulting curve consists of data pairs (input and output) that can be simply plotted on the S–N plane. Using the model in this way suffices for the subsequent analysis, but output data, even if this is not necessary, can be easily fitted by a 2nd to 4th order polynomial equation, as shown in Fig. 4.10. Compared to Fig. 4.9, symbol frequency has been reduced and the scale of the x-axis has been extended to $\log(N) = 2$ in Fig. 4.10, but only for a better presentation of the research findings.

4.2.3 Comparison of the Modeling Ability of the S–N Curve Formulations

The fatigue data from Chap. 2 were analyzed using the selected methods and the fatigue model parameters were determined and, for cases arbitrarily selected from those examined, are tabulated in Table 4.1 After the determination of the fatigue model parameters, S–N curves can be easily derived for any range of loading cycles.

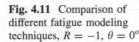

Fig. 4.11 Comparison of different fatigue modeling techniques, $R = -1$, $\theta = 0^o$

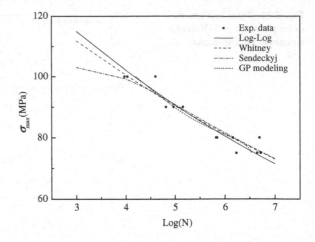

Fig. 4.12 Comparison of different fatigue modeling techniques, $R = 0.1$, $\theta = 15^o$

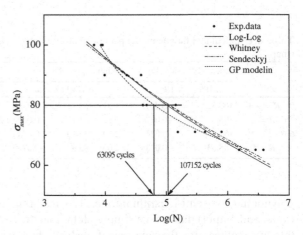

Fatigue behavior as modeled by all the available methods is presented in Figs. 4.11, 4.12, 4.13 and 4.14 for comparison. It can be concluded that although based on different approaches, in general all fatigue models could adequately represent the fatigue behavior of the selected experimental data, at least for the central part of the S–N curve, for $\log(N) = 3$ to $\log(N) = 6$. In all examined cases, Figs. 4.11, 4.12, 4.13 and 4.14 show that all fatigue models produce similar S–N curves with the Log–Log being steeper than the others. In some of the examined cases however, GP is shown to be superior as it can follow the real trend of the experimental data, without any constraints regarding a selected equation type. In Fig. 4.12, for example, where the predictions for the data obtained from specimens cut at 15° are presented, it is shown that the GP curve "follows" the trend of the experimental data more accurately than the other three fatigue models that result in a rather straight curve on the $\log(N)$-S plane.

For example, when examining the stress level of 80 MPa, the experimental average number of cycles could be calculated as 77,985 and the corresponding

Fig. 4.13 Comparison of different fatigue modeling techniques, $R = 10$, $\theta = 60^o$

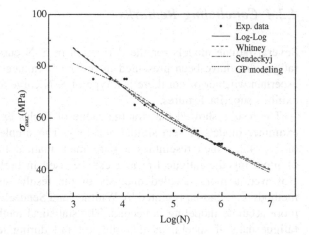

Fig. 4.14 Comparison of different fatigue modeling techniques, $R = 0.1$, $\theta = 75^o$

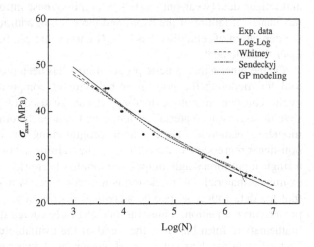

estimated numbers from the GP curve and other methods as 63,095 and 107,152 cycles, respectively. Moreover, for the stress level of 71 MPa, the GP tool estimates 380,189 loading cycles and the other methods approximately 562,341 loading cycles, while the experimental average is 421,213 loading cycles. For both examined stress levels, the GP curve underestimates the actual number of loading cycles by a factor of 9.7–19.1%, while the other curves overestimate the number of loading cycles by a factor of more than 33.5%. Although this conclusion is specific to the examined dataset, it shows the ability of the GP to adapt the model output to the given dataset.

The above comments were based on analysis of the available experimental fatigue data, without considering the corresponding static strengths of the examined material. Therefore, the validity of the modeling, without the need for extrapolation, is true only for numbers of cycles between ca 1,000 and 10 million.

4.2.4 Concluding Remarks

Several fatigue models for the derivation of S–N curves based on simple input fatigue data have been presented in order to evaluate their modeling ability and ascertain whether or not there is a type of S–N curve among those examined that exhibits superior features.

The results showed that for the range of cycles examined here (10^3-10^7) all examined models present similar accuracy. The simple deterministic formulation of Eq. 4.2, which resembles a pure mathematical fitting equation, is capable of modeling the fatigue life and can be used in preliminary design processes. However, a more detailed analysis of the results suggests that the statistical methods (e.g., those presented by Whitney and Sendeckyj) must be preferred when more accurate modeling is needed. The statistical methods take into account the fatigue data of specimens that did not fail during loading (run-outs) and are moreover based on assumptions regarding the relationship between static strength and fatigue data (wear-out model), which in a sense introduces into the process the mechanics of failure of the examined material. In addition they can be applied for the derivation of reliability-based S–N curves that can be very useful in the design of critical elements.

On the other hand, genetic programming has been proved to be a very powerful tool for modeling the non-linear behavior of composite laminates subjected to cyclic constant amplitude loading. It can be used to model the fatigue life of several composite material systems, and can be favorably compared with other modeling techniques. Here genetic programming has been used as a stochastic non-linear regression analysis tool. As the training dataset has been structured with a single input for a single output, the tool has been used to "fit" the behavior of the examined material. GP modeling is not based on any assumptions, as for example that the data follow a specific statistical distribution, or that the S–N curve is a power curve equation. Thus, the derived S–N curves do not follow any specific mathematical form but only the trend of the available data, giving each time the best estimate for their behavior. However, as shown here, the output data can be easily fitted by simple 3rd to 4th order polynomial equations. Although these facts alone suffice to justify the effectiveness of genetic programming as a modeling tool for the fatigue analysis of composite materials, the idea of using this tool for these types of analyses is not restricted to this application. The objective is to use more complex genetic programming configurations by introducing models with multiple inputs, e.g., stress amplitude, maximum stress, stress ratio and off-axis angle, and attempt to assign the corresponding number of cycles to failure to each set of inputs. It is anticipated that in such an analysis GP can be used as a predicting tool for several purposes, such as off-axis fatigue life prediction and construction of constant life diagrams .

Based on the above conclusions, all the examined S–N curve types are appropriate for modeling the fatigue life of composite materials. The empirical S–N formulations, like the one presented in Eq. 4.2, can be used for any preliminary stage, since the model parameters can be estimated even by hand calculations. The more complicated numerical modeling represented here by the GP tool is useful when more a detailed description of fatigue behavior is required. The statistical methods Eqs. 4.3 and 4.4 are able to derive S–N curves for any desired reliability level and are therefore useful for design purposes where high reliability levels are desirable.

4.3 Constant Life Diagrams

A strong mean stress effect on the fatigue life of the on- and off-axis specimens was observed (see Chap. 2). The fatigue behavior of the examined material under different stress ratios was visualized in Chap. 2 by means of the constant life diagrams (see Figs. 2.11–2.13). In these figures, a linear interpolation between the known values was used, although, as will be shown in the following, this is not repeatedly the most appropriate method for composite materials.

In previous papers presented in the literature [27–32] it has been proved that, although the classic linear Goodman diagram is the most commonly used, particularly for metals, it is not suitable for composite materials, mainly because of the variation in the tensile and compressive strengths that they exhibit. The damage mechanisms under tension are different from those under compression. In tension, the composite material properties are generally governed by the fibers, while in compression the properties are mainly determined by the matrix and matrix-fiber interaction. Therefore, straight lines connecting the ultimate tensile stress (UTS) and the ultimate compressive stress (UCS) with points on the $R = -1$ line for different numbers of cycles are not capable of describing the actual fatigue behavior of composite materials. A typical constant life diagram (CLD) for composite materials is thus usually shifted to the right-hand side and the highest point located away from the line corresponding to zero mean stress, $\sigma_m = 0$.

Several CLD models have been presented in the literature [28–39] in order to deal with the aforementioned characteristics of composite materials. A comprehensive evaluation of the fatigue life predicting ability of the most commonly used and most recent is presented in [27]. Since the introduction of the constant life diagram concept by Gerber and Goodman back in the 19th century [40, 41], all presented methods have two common features–they represent the fatigue data on the (σ_m–σ_a)-plane and their formulation is based on the fitting of available fatigue data for specified R-ratios or the interpolation between them. The same concept has been followed for the derivation of CLDs for composite materials.

Starting from the basic idea of the symmetric and linear Goodman diagram and the non-linear Gerber equation, different modifications were proposed to cover the peculiarities of the behavior of composite materials. A linear model representing a modified Goodman diagram was presented in [42]. It is based on a single experimentally derived S–N curve and linear interpolation for the estimation of any others. More sophisticated models, although still based on the linear interpolation between known S–N curve values and static strength data, were presented and analytical expressions for the theoretical derivation of any desired S–N curve were developed based on this idea [33, 43]. In the proposed models, a minimum of amount of experimental data was used, while simultaneously accommodating the particular characteristics of composites.

An alternative semi-empirical formulation was proposed in a series of papers by Harris's group [28, 29, 34]. The solution was based on fitting the entire set of experimental data with a non-linear equation to form a continuous bell-shaped line from the ultimate tensile stress to the ultimate compressive stress of the examined material. The drawback of this idea was the need to adjust a number of parameters based on experience and existing fatigue data. Kawai [30, 35] proposed the so-called anisomorphic CLD that can be derived by using only one "critical" S–N curve. The critical R-ratio is equal to the ratio of the ultimate compressive over the ultimate tensile stress of the examined material. The obvious drawback of this model is the need for experimental data for this specific S–N curve and therefore, theoretically, it cannot be applied to existing fatigue databases. However, the minimum amount of data required is an asset of the proposed methodology. Based on the Gerber line, another formulation of the CLD was proposed by Boerstra [36]. This offers a simple method for the lifetime prediction of laminated structures subjected to fatigue loads with continuously varying mean stress and dispenses with any classification of fatigue data according to R-values. The disadvantage of this method is the complicated optimization process with five variables that must be followed in order to derive the CLD model.

A new model was recently proposed by Kassapoglou [37, 38] based on the assumption that the probability of failure during any fatigue cycle is constant and equal to the probability of failure under static loading. Following this assumption, S–N curves under any loading pattern can be derived by using only tensile and compressive static strength data. However, the restricted use of static data disregards the different damage mechanisms that develops during fatigue loading and in many cases leads to erroneous results, e.g., [44].

A novel constant life diagram formulation was introduced in [39]. The model was established on the basis of the relationship between the stress ratio (R) and the stress amplitude (σ_a). Simple phenomenological equations were derived from this relationship without the need for any assumptions. The model parameters can be estimated based on a limited number of fatigue data. The validity of the proposed CLD formulation was evaluated by comparing predicted and experimental results for a wide range of composite materials. This new formulation, designated the "Piecewise Non-Linear model" (PNL), compares well with the existing ones, being more accurate in some of the studied cases for a wide range of glass and carbon fiber composite materials [39].

Novel computational methods have also been employed during the last decade for modeling the fatigue behavior of composite materials and the derivation of constant life diagrams based on limited amounts of experimental data, e.g., [8, 18, 19]. These methods offer a means of representing the fatigue behavior of the examined composite materials that is not biased by any damage mechanisms and not restricted by any mathematical model description. They are data-driven techniques and their modeling quality depends on the quality of the available experimental data.

The influence of the constant life diagram formulation on the prediction of the fatigue life of composite materials was extensively studied in [27]. The most commonly used and most recent CLD formulations for composite materials are evaluated in this paragraph. The applicability of the models, the need for experimental data and the accuracy of their predictions are considered critical parameters for the evaluation. The effect of the selection of the CLD formulation on fatigue life prediction is assessed according to its ability to accurately estimate unknown S–N curves. The comparison of the modeling ability of the CLD formulations can also be based on the life prediction results of any life prediction methodology of which the CLDs are part. In this case however, other parameters that influence the results (i.e., other steps of the methodology concerning the selection of the S–N curve type for the data interpretation, damage summation rule, etc.) may mask the effect of the CLD formulation. Based on the results, recommendations concerning the applicability, advantages and disadvantages of each of the examined CLD formulations are discussed.

4.3.1 Theory of CLD Models

Constant life diagrams reflect the combined effect of mean stress and material anisotropy on the fatigue life of the examined composite material. Furthermore, they offer a predictive tool for the estimation of the fatigue life of the material under loading patterns for which no experimental data exist. The main parameters that define a CLD are the mean cyclic stress, σ_m, the cyclic stress amplitude, σ_a, and the R-ratio defined as the ratio between the minimum and maximum cyclic stress, $R = \sigma_{min}/\sigma_{max}$. A typical CLD annotation is presented in Fig. 4.15.

As shown, the positive $(\sigma_m$–$\sigma_a)$-half-plane is divided into three sectors, the central one comprising combined tensile and compressive loading. The Tension-Tension (T-T) sector is bounded by the radial lines, representing the S–N curves at $R = 1$ and $R = 0$, the former corresponding to static fatigue and the latter to tensile cycling with $\sigma_{min} = 0$. S–N curves belonging to this sector have positive R-values less than unity. Similar comments regarding the two remaining sectors can be derived from the annotations shown in Fig. 4.15. Every radial line with $0 < R < 1$, i.e., in the T-T sector, has a corresponding symmetric line with respect to the σ_a-axis, which lies in the compression-compression (C-C) sector and whose R-value is the inverse of the tensile R-value, e.g., $R = 0.1$ and $R = 10$.

Radial lines emanating from the origin are expressed by:

$$\sigma_a = \left(\frac{1-R}{1+R}\right)\sigma_m \tag{4.5}$$

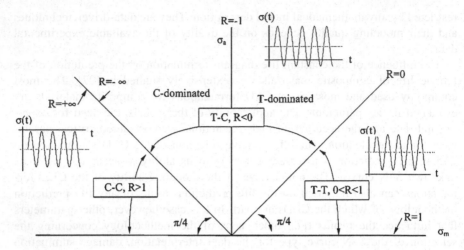

Fig. 4.15 Annotation for $(\sigma_m - \sigma_a)$-plane

and represent single S–N curves. Data on these lines belong to the S–N curve for that particular stress ratio. Constant life diagrams are formed by joining data points corresponding to the same numbers of cycles on consecutive radial lines.

Although from a theoretical point of view the above representation of the CLD is rational, it presents a deficiency, since it cannot accurately model the fatigue behavior of the examined material for loadings in the regions of the T-T (e.g., $R = 0.95$) and C-C (e.g., $R = 1.05$) sectors close to the horizontal axis, which represent loading under very low stress amplitude and high mean values with a culmination for zero stress amplitude ($R = 1$).

The classic CLD formulations require that the constant life lines converge to the ultimate tensile stress (UTS) and the ultimate compressive stress (UCS), regardless of the number of loading cycles. However, this is an arbitrary simplification originating from the lack of information about the fatigue behavior of the material when no amplitude is applied. In fact, this type of loading cannot be considered fatigue loading, but rather creep of the material (constant static load over a short or long period). Although modifications that take the time-dependent material strength into account have been introduced, their integration into CLD formulations requires the adoption of additional assumptions, see e.g., [32, 44].

4.3.1.1 Linear CLD

The concept of the linear CLD model [42] is based on a single S–N curve that must be experimentally derived. All other S–N curves can be determined from the given one by simple calculations. This simplified formulation assumes that the failure mechanism is identical in tension and in compression when the load amplitude is the same. In the $(\sigma_m - \sigma_a)$-plane, the above assumption implies that any constant life

Fig. 4.16 Linear CLD for
on-axis specimens,
$N = 10^3$–10^7

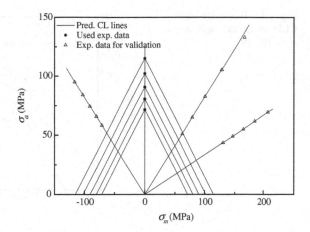

line forms an isosceles triangle, subtending $\pi/4$ angles with the axes [42]. Any
constant life line can be calculated by:

$$\frac{\sigma_a}{\sigma_o} + \frac{\sigma_m}{\sigma_o} = N^{-1/k} \tag{4.6}$$

where k and σ_0 are parameters of the power law equation which describes the S–N
curve at the selected R-value.

The linear CLD for on-axis specimens of the examined material in Chap. 2 is
presented in Fig. 4.16, based on the experimentally derived S–N curve under
reversed loading.

4.3.1.2 Piecewise Linear CLD

The piecewise linear CLD [33, 43] is derived by linear interpolation between
known values in the $(\sigma_m\text{–}\sigma_a)$-plane. This CLD model requires a limited number of
experimentally determined S–N curves along with the ultimate tensile and compressive stresses of the materials. S–N curves representing the entire range of
possible loading are commonly used for the construction of piecewise linear
CLDs, normally at $R = 0.1$ for T-T loading, $R = -1$ for T-C loading and $R = 10$
for C-C loading patterns. Constant life lines connect data points of the same
number of cycles on various S–N curves. Unknown S–N curves are calculated by
linear interpolation between known values of fatigue and static strength data.

Analytical expressions were developed for the description of each region of the
piecewise linear CLD in [43].

1. If R' is in the T-T sector of the CLD, and between $R = 1$ and the first known
   R-ratio on the $(\sigma_m\text{-}\sigma_a)$-plane when moving counterclockwise, R_{1TT}, then

Fig. 4.17 Piecewise linear
CLD for on–axis specimens,
$N = 10^3$–10^7

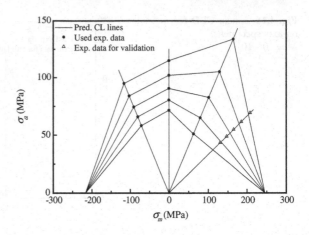

in which σ'_a and $\sigma_{a,1TT}$ are the stress amplitudes corresponding to R' and R_{1TT},
respectively and $r_i = (1 + R_i)/(1 - R_i)$, and $r' = (1 + R')/(1 - R')$.

2. If R' is located between any of two known R-ratios, R_i and R_{i+1},

$$\sigma'_a = \frac{\sigma_{a,i}(r_i - r_{i+1})}{(r_i - r')\frac{\sigma_{a,i}}{\sigma_{a,i+1}} + (r' - r_{i+1})} \tag{4.8}$$

3. If R' lies in the C-C region of the CLD, and between $R = 1$ and the first known
   R-ratio in the compression region, R_{1CC},

$$\sigma'_a = \frac{UCS}{\frac{UCS}{\sigma_{a,1CC}} - r' + r_{1CC}} \tag{4.9}$$

where σ'_a and $\sigma_{a,1CC}$ are the stress amplitudes corresponding to R' and R_{1CC},
respectively.

The application of Eqs. 4.7–4.9 to the material data presented in Chap. 2 results
in the CLD shown in Fig. 4.17 for the on-axis specimens.

4.3.1.3 Harris's CLD

Harris and his colleagues developed a semi-empirical equation based on fatigue
test data obtained from a range of carbon and glass fiber composites [28, 29, 34]

$$a = f(1 - m)^u(c + m)^v \tag{4.10}$$

where a is the normalized stress amplitude component, σ_a/UTS, m the normalized mean stress component, σ_m/UTS, and c the normalized compression strength, UCS/UTS. In this equation, f, u and v are three adjustable parameters that are functions of fatigue life. The estimation of the parameters is based on the behavior of the material along the entire range of loading, combining T-T, C-C and T-C fatigue results. In this context, the formulation takes into account the combined effect of the different failure mechanisms that are developed under tension and under compression on the fatigue life. Early studies [28, 29] showed that parameter f mainly controls the height of the curve, and is a function of the ratio of the compressive to the tensile strength, while the exponents u and v determine the shapes of the two 'wings' of the bell-shaped curve. Initially, the model was established with two simplified forms of Eq. 4.10 where $u = v = 1$ and $u = v$ for a family of carbon/Kevlar unidirectional hybrid composites [28, 29]. Since this model was not accurate for different material systems, the general form of Harris's model was implemented in the sequel. In the general form, parameters f, u and v were considered as functions of fatigue life. Depending on the examined material, and the quality of the fatigue data, these parameters were found to depend linearly on the logarithm of fatigue life, $\log(N)$, for a wide range of FRP materials [29]:

$$f = A_1 \log(N) + B_1 \tag{4.11}$$

$$u = A_2 \log(N) + B_2 \tag{4.12}$$

$$v = A_3 \log(N) + B_3 \tag{4.13}$$

in which parameters A_i and B_i, $i = 1,2,3$, are determined by fitting Eqs. 4.11–4.13 to the available experimental data for different loading cycles.

Beheshty and Harris [29] showed that the selection of this empirical form for the parameters u and v can be employed for a wide range of materials, especially CFRP laminates. However, parameter f is sensitive to the examined material and its values vary considerably between GFRP and CFRP laminates. Since the modeling accuracy of the Harris CLD is significantly dependent on the quality of the fitting of these parameters, Harris and his colleagues established different formulations for the estimation of parameter f based on experimental evidence obtained from a number of different composite material systems. The most recent proposal for the estimation of parameter f is the following equation:

$$f = Ac^{-p} \tag{4.14}$$

where A and p are also functions of $\log(N)$. However, experimental evidence proved that values of $A = 0.71$ and $p = 1.05$ can be used in order to produce acceptable results for a wide range of CFRP and GFRP laminates [34].

The Harris CLD, also referred as the bell-shaped CLD, e.g., [42], looks like the one presented in Fig. 4.18 for the on-axis specimens of the materials of the

Fig. 4.18 Harris CLD for
on-axis specimens,
$N = 10^3 - 10^7$

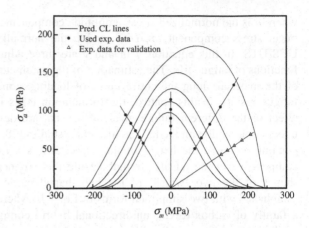

examined dataset. Equation 4.14 with $A = 0.71$ and $p = 1.05$ was used for the
derivation of parameter f. The other two parameters were estimated by fitting
Eqs. 4.12–4.13 to the experimental data.

4.3.1.4 Kawai's CLD

Kawai's group [30, 35] developed a formula that describes an asymmetric constant
life diagram, designated the anisomorphic constant fatigue life (CFL) diagram in
[30]. The basic characteristic of this formulation is that it can be constructed by
using only one experimentally derived S–N, designated the critical S–N curve. The
R-ratio of this S–N curve is defined as the ratio of the ultimate compressive over
the ultimate tensile stress of the examined material. The formulation is based on
three main assumptions:

1. The stress amplitude, σ_a, for a given constant value of fatigue life N is greatest
   at the critical stress ratio,
2. The shape of the CFL curves changes progressively from a straight line to a
   parabola with increasing fatigue life, and
3. The diagram is bounded by the static failure envelope, i.e., two straight lines
   connecting the ultimate tensile and ultimate compressive stresses with the
   maximum σ_a on the critical S–N curve.

The CFL formulation depends on the position of the mean stress on the
$(\sigma_m - \sigma_a)$-plane, whether it is in the tensile or compressive region. The mathematical formulation reads:

$$\frac{\sigma_a^\chi - \sigma_a}{\sigma_a^\chi} = \begin{cases} \left(\frac{\sigma_m - \sigma_m^\chi}{UTS - \sigma_m^\chi}\right)^{(2-\psi_x)}, & UTS \geq \sigma_m \geq \sigma_m^\chi \\ \left(\frac{\sigma_m - \sigma_m^\chi}{UCS - \sigma_m^\chi}\right)^{(2-\psi_x)}, & UCS \leq \sigma_m \leq \sigma_m^\chi \end{cases} \qquad (4.15)$$

Fig. 4.19 Kawai's CLD for on-axis specimens, $N = 10^3–10^7$

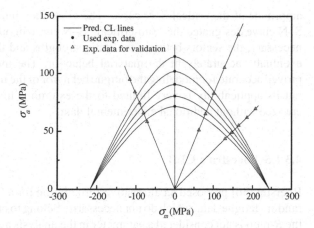

where σ_m^χ and σ_a^χ represent the mean and cyclic stress amplitudes for a given constant value of life N under fatigue loading at the critical stress ratio. ψ_χ denotes the critical fatigue strength ratio and is defined as:

$$\psi_\chi = \frac{\sigma_{max}^\chi}{\sigma_B} \quad (4.16)$$

where σ_{max}^χ is the maximum fatigue stress for a given constant value of life N under fatigue loading at the critical stress ratio. $\sigma_B(>0)$ is the reference strength (the absolute maximum between UTS and UCS) of the material that defines the peak of the static failure envelope. Therefore this normalization guarantees that ψ_χ always varies in the range [0, 1] and the exponents $(2 - \psi_\chi)$ in Eq. 4.15 are always greater than unity. Subsequently, linear (when $2 - \psi_\chi = 1$) or parabolic (when $2 - \psi_\chi > 1$) curves can be obtained from Eq. 4.15.

The critical fatigue strength ratio (see Eq. 4.16) is related to the number of loading cycles defining the normalized critical S–N curve:

$$\psi_\chi = f(2N_f) \quad (4.17)$$

After determining the critical S–N curve by fitting to the available fatigue data, the CFL diagram can be constructed on the basis of the static strengths, UTS and UCS, and the critical S–N relationship. The dependence of the anisomorphic CFL diagram on the critical S–N curve limits its applicability to the datasets for which available data under the critical R-ratio exist. When no-S–N curve under the critical R-ratio exists, the one that is closest to this value can be used as proposed in [27]. This was also the process followed for the derivation of the anisomorphic CFL diagram of the on-axis specimens of the examined dataset, which is presented in Fig. 4.19.

A modified anisomorphic CFL formulation was recently introduced by Kawai and Murata [45] to improve the performance of the original anisomorphic diagram. The authors observed that the fatigue behavior of matrix-dominated CFRP laminates cannot be described by simple linear or curved lines between the UTS, and the stress amplitude of the critical S–N curve. Therefore they introduced the use of another S–N curve designated the "sub-critical" S–N curve in order to subdivide, wherever necessary, the sectors between the static strengths and the critical S–N curve, and eventually accurately fit the material behavior. The modified CFL diagram was proved accurate for modeling the fatigue behavior of the material investigated in [45] but its applicability is very limited to the examined material and cannot be generalized without additional experimental data.

4.3.1.5 Boerstra's CLD

Boerstra [36] proposed an alternative formulation for a CLD that can be applied to random fatigue data, which do not necessarily belong to an S–N curve. In this model, the R-ratio is not considered a parameter in the analysis and the model can be applied to describe the behavior of the examined material under loads with continuously changing mean and amplitude values. Boerstra's model constitutes a modification of the Gerber line. The exponent was replaced by a variable also including the difference in tension and compression. The general formulae of the model are:

$$\text{For } \sigma_m > 0 : \quad \sigma_{ap} = \sigma_{Ap}(1 - (\sigma_m/UTS)^{\alpha T}) \tag{4.18}$$

$$\text{For } \sigma_m < 0 : \quad \sigma_{ap} = \sigma_{Ap}(1 - (\sigma_m/UCS)^{\alpha C}) \tag{4.19}$$

where σ_{ap} is the stress amplitude component for a reference number of cycles, N_p, σ_{AP} is an "apex" stress amplitude for N_p and $\sigma_m = 0$, and αT and αC are two shape parameters of the CLD curves for the tensile and compressive sides, respectively.

The above equations represent the CLD lines in the $(\sigma_m$–$\sigma_a)$-plane. According to the author [35], existing fatigue data for different kinds of composite materials show steeper S–N curves under tension and than under compression. An exponential relationship with the mean stress can be a good description for the slope $(1/m)$ of S–N lines as follows:

$$m = m_o e^{(-\sigma_m/D)} \tag{4.20}$$

in which m_o is a measure for the slope of the S–N curve on the Log–Log scale for $\sigma_m = 0$ and D is the skewness parameter for the dependency of m.

Equations 4.18–4.20 suggest that five parameters, m_o, D, N_p, αT, and αC, must be defined in order to construct the CLD model. However, the estimation of the parameters requires a multi-objective optimization process. The aim of this optimization is to estimate the parameters allowing the calculation of the shortest distance between each measuring point and the S–N line for its particular mean stress. The procedure is as follows:

1. The static strengths UTS and UCS are determined and some fatigue test data on specimens with various values of stress amplitude, σ_α, and mean stress, σ_m, should also be available.

Fig. 4.20 Boerstra CLD for on-axis specimens, based on experimental data, $N = 10^3 – 10^7$

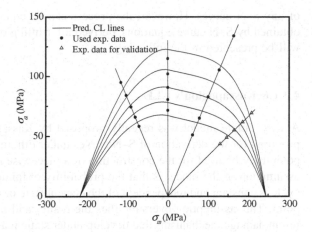

2. The desired value of N_p is chosen and an initial set of values for parameters m_o, D, N_p, αT and αC is assumed.
3. The slope of the S–N line, m, is calculated for each measured σ_m using Eq. 4.20.
4. The σ_α corresponding to each σ_m is projected to the $(\sigma_m – \sigma_a)$-plane for the selected number of cycles N_p by $\sigma_{ap} = \sigma_a (N / N_p)^{(1/m)}$.
5. $\sigma_{\alpha P}$ is calculated for each pair of σ_{0p} and σ_m using Eqs. 4.18 and 4.19.
6. A modified stress amplitude, $\sigma_{ap,mod}$, is calculated by feeding back the average value of σ_{AP} and the measured mean stress value, σ_m, into Eqs. 4.18 and 4.19.
7. The difference between the logarithms of the measured stress amplitude and the modified stress amplitude is then computed as: $\Delta\sigma_a = \ln(\sigma_{ap}) - \ln(\sigma_{ap,mod})$.
8. The theoretical number of cycles, N_e, corresponding to the $\sigma_{ap,mod}$ stress amplitude and the measured mean stress, σ_m, can be calculated by solving the equation:

$$N_e = N_p \left(\frac{\sigma_{a,mod}}{\sigma_a} \right)^m \tag{4.21}$$

9. The difference between the measured number of cycles, N, and the theoretical number of cycles, N_e, is defined by $\Delta n = \ln(N) - \ln(N_e)$.
10. The shortest distance between each independent point and the S–N lines in the $\sigma_m – \sigma_a – N$ space is expressed by: $\Delta t = sign(\Delta\sigma_a)\sqrt{(1/(1/\Delta\sigma_a^2 + 1/\Delta n^2))}$. The sum of all Δts is designated as the total standard deviation, SDt. Minimization of the SDt results in the estimation of the optimal m_o, D, N_p, αT and αC parameters.

The CLD that results from the above is presented in Fig. 4.20. The experimental fatigue data were directly considered in the analysis without the derivation of any S–N curves. However, the process can be equally applied to fatigue data obtained by S–N curve equations, derived after fitting of the experimental data, as will be presented in the following.

4.3.1.6 Kassapoglou's CLD

A very simple model was recently proposed by Kassapoglou [37, 38]. Although proposed for the derivation of S–N curves under different R-ratios, this model can potentially be used for the construction of a piecewise non-linear CLD. The basic assumption of the model is that the probability of failure of the material during a cycle is constant and independent of the current state or number of cycles up to this point. This assumption oversimplifies the reality and masks the effect of the different damage mechanisms that develop under static loading and at different stages of fatigue loading. However, adoption of this assumption allows the estimation of the parameters of a single distribution based on the static strength data and use of this same distribution for the calculation of the fatigue life of the examined material. In this case, this model requires no fatigue testing, no empirically determined parameters and no detailed modeling of damage mechanisms.

The model comprises the following equations for calculation of maximum cyclic stress as a function of number of cycles:

$$\sigma_{max} = \frac{\beta_T}{(N)^{\frac{1}{a_T}}}, \text{ for } 0 \leq R < 1 \tag{4.22}$$

$$\sigma_{max} = \frac{\beta_C}{(N)^{\frac{1}{a_C}}}, \text{ for } R > 1 \tag{4.23}$$

while for $R < 0$ the following equation should be solved numerically:

$$N = \frac{1}{\left(\frac{\sigma_{max}}{\beta_T}\right)^{a_T} + \left(\frac{\sigma_{min}}{\beta_C}\right)^{a_C}} \tag{4.24}$$

Parameters a_i, β_i, $i = T$ or C denote the scale and shape of a two-parameter Weibull distribution that can describe the static data in tension and compression respectively.

This model cannot be applied to the examined material described in Chap. 2 since the amount of available experimental static strength data is limited and therefore it is not possible to fit a reliable statistical distribution on them.

4.3.1.7 The Piecewise Non-Linear CLD

All previous CLD formulations are based on the fitting of linear or non-linear equations to existing fatigue data on the $(\sigma_m-\sigma_a)$-plane. However, there is no

Fig. 4.21 Representation of
relationship between fatigue
parameters σ_a–R–$\text{Log}(N)$

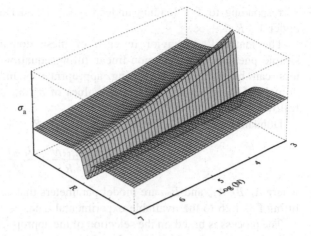

Fig. 4.22 Representation of
constant life diagram on R–σ_a
plane

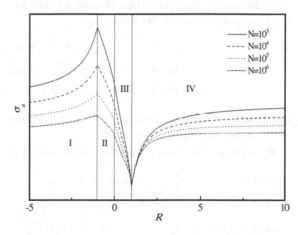

rational explanation for the selection of these two stress parameters. Any other combination of σ_a–σ_m–R can just as well be used for the derivation of a constant life diagram. A plot of stress amplitude against stress ratio for different numbers of loading cycles is presented in Fig. 4.21.

The surface of Fig. 4.21 represents the fatigue failure locus of the examined material. Any loading combination above the surface causes failure. A projection of this surface on the R–σ_a plane can be considered as a constant life diagram, see Fig. 4.22.

In Fig. 4.22, the x-axis represents the R-ratio and ranges from $-\infty$ to $+\infty$ without any singularity. The y-axis represents the stress amplitude and has positive values. S–N curves for any stress ratio, R, are represented by vertical lines emanating from the corresponding value of R on the x-axis. This diagram can be divided into four distinct domains, each corresponding to different loading conditions: Domain I for compression-tension (C-T) loading under $-\infty \leq R \leq -1$, Domain II for tension-compression (T-C) loading under $-1 \leq R \leq 0$, Domain III corresponding to T-T loading under $0 \leq R \leq 1$ and Domain IV for C-C loading under $1 \leq R \leq +\infty$.

The material's behavior in each of these domains can be described by simple phenomenological non-linear fitting equations and the model parameters can be estimated by using appropriate boundary conditions for each domain of the diagram and known values of σ_a, σ_m and R, as described in the following:

Domains *I and IV*: $-\infty \leq R \leq -1$ and $1 \leq R \leq +\infty$:

$$\sigma_a = (1 - R)\left(\frac{A_{I, \text{or } IV}}{R} + \frac{B_{I, \text{or } IV}}{R^2}\right) \tag{4.25}$$

where A_I, B_I, A_{IV} and B_{IV} are model parameters that can be easily determined by fitting Eq. 4.26 to the available experimental data.

The process is based on the selection of the appropriate boundary conditions for each domain of the CLD. For Domains I and IV, described by Eq. 4.25, the boundary conditions are the following:

$$\text{for } R = -1, \ \sigma_a = \sigma_a^{R=-1}, \ \text{and } \sigma_m = 0,$$
$$\text{for } R = \pm\infty, \ \sigma_a = \sigma_a^{R=\pm\infty}, \ \text{and} \tag{4.26}$$
$$\text{for } R = 1, \ \sigma_a = 0, \ \text{and } \sigma_m = UCS$$

where stress parameter superscripts denote the corresponding stress ratio, e.g. $\sigma_\alpha^{R=-1}$ is the stress amplitude for $R = -1$.

By applying these three boundary conditions, Eq. 4.25 becomes:

$$\text{for } R = -1 \ \rightarrow \ \sigma_a^{R=-1} = 2(-A_I + B_I), \ \text{or}$$
$$\text{for } R = 1 \ \rightarrow \ UCS = 2(A_{IV} + B_{IV}), \ \text{or} \tag{4.27}$$
$$\text{for } R = \pm\infty \ \rightarrow \ \sigma_a^{R=\pm\infty} = \lim_{R \to \pm\infty}(1 - R)\left(\frac{A_{I, \text{or } IV}}{R} + \frac{B_{I, \text{or } IV}}{R^2}\right) = -A_{I, \text{or } IV}$$

and the four fitting parameters A_I, B_I, A_{IV} and B_{IV} can be defined as:

$$A_{I, \text{or } IV} = -\sigma_a^{R=\pm\infty}$$
$$B_I = \frac{\sigma_a^{R=-1}}{2} - \sigma_a^{R=\pm\infty} \tag{4.28}$$
$$B_{IV} = \frac{UCS}{2} + \sigma_a^{R=\pm\infty}$$

When the S–N curve under $R = 10$ is available instead of that under $R = \pm\infty$, the boundary conditions should be adjusted accordingly.

Domains II and III: $-1 \leq R \leq 0$ and $0 \leq R \leq 1$:

$$\sigma_a = \frac{1 - R}{A_{II, \text{or } III} R^n + B_{II, \text{or } III}} \tag{4.29}$$

The fitting of Eq. 4.29 on the constant amplitude fatigue data of several different material systems [39] proved that parameter n can be considered equal to 1 for Domain II and equal to 3 for Domain III. The boundary conditions are the following:

$$\text{for } R = 1, \ \sigma_a = 0 \text{ and } \sigma_m = UTS$$
$$\text{for } R = -1, \ \sigma_a = \sigma_a^{R=-1} \text{ and } \sigma_m = 0 \tag{4.30}$$

if only the S–N curves under $R = -1$ and $R = 1$ are used. Implementing the above-mentioned boundary conditions results in:

$$A_{II,\text{or }III} = \frac{1}{UTS} - \frac{1}{\sigma_a^{R=-1}}$$
$$B_{II,\text{or }III} = \frac{1}{UTS} + \frac{1}{\sigma_a^{R=-1}} \tag{4.31}$$

However, if the S–N curve under $R = 0$ is considered as well, the boundary conditions Eq. 4.31 are supplemented by:

$$\sigma_a = \sigma_a^{R=0}, \ \text{for } R = 0 \tag{4.32}$$

By applying the boundary conditions for Domains II and III in Eq. 4.29, parameters A_{II}, B_{II}, A_{III} and B_{III} acquire the following values:

$$A_{II} = \frac{1}{\sigma_a^{R=0}} - \frac{2}{\sigma_a^{R=-1}}$$
$$A_{III} = \frac{2}{UTS} - \frac{1}{\sigma_a^{R=0}} \tag{4.33}$$
$$B_{II,\text{or }III} = \frac{1}{\sigma_a^{R=0}}$$

Similarly to Domains I and IV, when the S–N curve under $R = 0.1$ is available instead of that under $R = 0$, the boundary conditions are modified accordingly. More S–N curves may be used to improve the accuracy of the model. However, as shown in the next paragraphs, the use of only two or three S–N curves, under $R = -1$, $R = \pm\infty$ (alternatively $R = 10$), and $R = 0$ (alternatively $R = 0.1$) suffices to produce an accurate model.

The PNL constant life diagram for the on-axis specimens on the $(\sigma_m - \sigma_a)$-plane is presented in Fig. 4.23. The S–N curves derived under $R = 10$, -1 and 0.1 have been used for the calibration of the model.

4.3.2 Evaluation of the CLD Models

The performance of the examined constant life diagrams has been evaluated on three different material systems. In addition to the material examined in this book

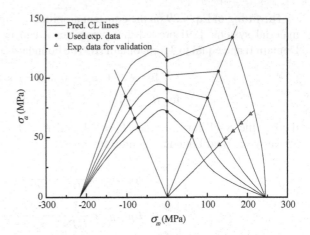

Fig. 4.23 PNL CLD for on-axis specimens, $N = 10^3$–10^7

(see Chap. 2), two more material databases were used in order to support the analysis and assist the derivation of more reliable conclusions. All examined materials are fiberglass-polyester and fiberglass-epoxy laminates, which are typical materials used in the wind turbine rotor blade construction industry.

The following criteria were considered to evaluate the applicability of the examined CLD models and assess their influence on the fatigue life prediction of the examined composite materials:

Accuracy of predictions: quantified by the accuracy of predicting new S–N curves. Need for experimental data: quantified by the number of S–N curves required to apply each CLD model.
Difficulty of application: qualitative criterion.
Implemented assumptions: qualitative criterion.

For the application of the linear model, the $R = -1$ curve was used. For the construction of the Kawai CFL, the $R = -1$ curve (the closest to the critical one, since the S–N curves corresponding to the critical R-ratios (ca. -0.9 for the on-axis specimens and ca. -0.8 for the specimens cut at 45° off-axis) are not available) together with the static strengths were employed. In addition, the $R = -0.5$ curve was used for the modeling of material #3, since for this case the critical R-ratio was -0.63. For the application of the remaining models – piecewise linear, Harris and Boerstra—three to five S–N curves, under $R = 0.1$, $R = -1$, $R = 10$ and additionally $R = 0.5$ and $R = 2$ (for material #3) along with the static strengths were used to describe all the regions of the CLD. Kassapoglou's model was applied only for the database of material #3, since it was the only one for which a statistically significant population of static strength data was available. Power curve fitting was performed on all available experimental fatigue data to determine the S–N curves. These curves were used in all formulations in order to have the same basis for the comparisons.

Material #1 GFRP multidirectional specimens cut on-axis and at 45° off-axis from a laminate with the stacking sequence $[0/(\pm 45)_2/0]_T$ (from Chap. 2).

Specimens cut on-axis and at 45° from the multidirectional laminate from Chap. 2 were considered as the first example for the comparison of the CLD formulations. The selected test set consisted of 56 (for the on-axis) and 57 (for the off-axis) valid fatigue data points, distributed in four S–N curves per case (at ratios $R = 0.5, 0.1, -1$ and 10). This is a typical dataset containing experimental fatigue data for the initial steps of design processes. Details concerning the specified material, preparation and testing procedures can be found in Chap. 2 of this volume. The UTS and UCS for this material were experimentally determined by axial tests as being 244.84 MPa and 216.68 MPa for the on-axis and 139.12 MPa and 106.40 MPa for the 45° off-axis specimens.

Three of the four existing S–N curves and the static strength values were used as the input data. The S–N curve at $R = -1$ was used for the construction of the linear CLD and the Kawai CFL. For the model proposed by Boerstra, this is not necessary, as it can be applied even for sparse fatigue data in the $(\sigma_a-\sigma_m-N)$-space. However, the CLD based on the fitted S–N curves was also plotted. Pre-processing of the fatigue data revealed that Eq. 4.14 is more appropriate for determination of parameter f, as prescribed by the Harris model. Equations (4.12) and (4.13) were used for the u and v parameters. The resulting CLD based on the estimation of all parameters using Eqs. 4.11–4.13 was also derived for the comparisons.

Different CLDs based on the various approaches are presented in Figs. 4.16, 4.17, 4.18, 4.19, 4.20 and 4.23 for the on-axis specimens, and Fig. 4.24 for the 45° off-axis specimens. Based on the results, Linear and Kawai CLDs are inaccurate for the examined material, while the predictions of the piecewise linear and Boerstra diagrams at the stress ratio $R = 0.5$ seem the most accurate.

It should be mentioned that the S–N curve used as the critical one is different from that recommended by Kawai's model and this may be the reason for the inaccurate results. As shown in Fig. 4.24c, the influence of parameter f on the shape of the predicted CLD based on Harris's model is very important. Insufficient modeling of its relationship to fatigue life can introduce significant errors. However, use of both fitting equations, Eqs. 4.11–4.13 or Eq. 4.14, introduced errors, especially in the vicinity of the low-cycle fatigue region. The application of Boerstra's model based on the experimental results instead of the fitted S–N curves led to a less conservative CLD for the examined material. The accuracy is not significantly affected by this selection however.

Material #2 Multidirectional glass-polyester laminate with a stacking sequence that can be encoded as $[(\pm 45)_8/0_7]_S$, [46].

The second material used for the comparisons was a multidirectional glass-polyester laminate consisting of 50% per weight unidirectional and 50% per weight ±45 plies. The material data were initially produced for the FACT database and were subsequently included in the Optidat database [46]. A total of 101 valid fatigue data points were found for the predetermined material tested under

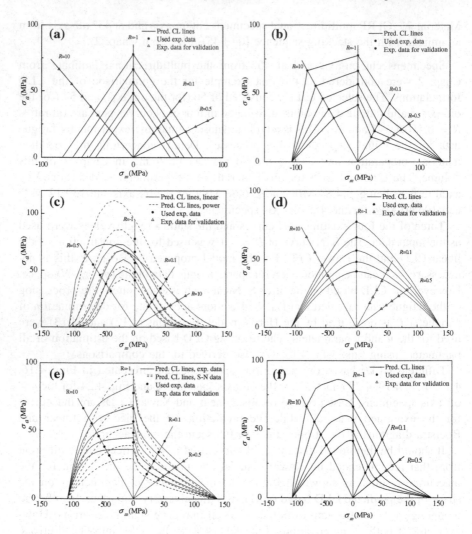

Fig. 4.24 Constant life diagrams for $N = 10^3–10^7$, material #1 (linear-a, piecewise linear-b, Harris-c, Kawai-d, Boerstra-e, piecewise non-linear-f)

five different constant amplitude conditions: $R = 0.1, −0.4, −1, 10, −2$ and used for comparisons. In this dataset, the maximum stress level ranged between 65 and 325 MPa and measured lifetime was between 48 for low-cycle fatigue, and 60.3 million cycles for longer lifetimes. A UTS of 370 MPa and UCS of 286 MPa were reported.

Three of the five existing S–N curves, those under $R = 10, −1$ and 0.1, plus the static strengths were used as input data. The remaining two S–N curves (under $R = −0.4$ and $−2$) were used to evaluate the modeling accuracy of the proposed methods. As for material #1, the $R = −1$ S–N curve was used for the construction

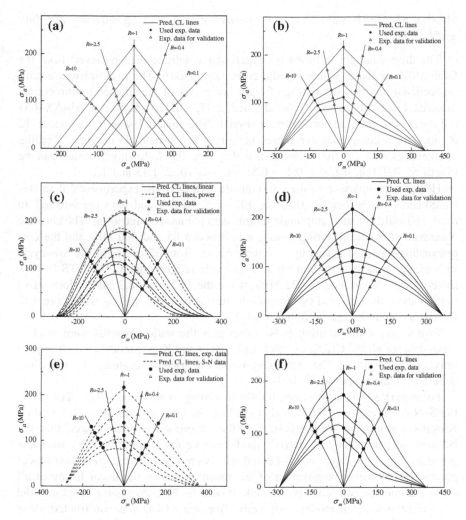

Fig. 4.25 Constant life diagrams for $N = 10^3–10^7$, material #2 (linear-a, piecewise linear-b, Harris-c, Kawai-d, Boerstra-e, piecewise non-linear-f)

of the linear and Kawai models, since it is the closest one to the critical stress ratio (−0.77). Based on the preprocessing of the experimental data, all model parameters for Harris were considered as linear functions of the log(N) and estimated by means of Eqs. 4.11–4.13.

Constant life diagrams according to the described models are presented in Fig. 4.25. All derived CLDs, except that prescribed by the linear model, are accurate for the prediction of the S–N curve at $R = −0.4$ and $R = −2.5$. As can be seen in Fig. 4.25e, the use of experimental data or fitted S–N data does not significantly affect the predictions based on Boerstra's model. The difference is even less when longer lifetime is evaluated.

Material #3   GFRP multidirectional laminate with a stacking sequence of [90/0/ ±45/0]$_S$, [47].

The third example is based on experimental fatigue data retrieved from the DOE/MSU database, which has the code name DD16. The material was a multidirectional laminate consisting of eight layers, six of the stitched unidirectional material D155 and two of the stitched, ±45, DB120. CoRezyn 63–AX–051 polyester was used as the matrix material. The material was tested under 12 R-ratios for a comprehensive representation of a constant life diagram. Reading counterclockwise on the constant life diagram, the following R-ratios can be identified: 0.9, 0.8, 0.7, 0.5, 0.1, −0.5, −1, −2, 10, 2, 1.43 and 1.1.

Here, for comparison of the constant life formulations, experimental data collected under seven R-ratios (0.8, 0.5, 0.1, −0.5, −1, −2 and 10) were selected. In total, 360 valid constant amplitude fatigue data points were retrieved. The absolute maximum stress level during testing was between 85 and 500 MPa and the corresponding recorded cycles up to failure ranged from 37 cycles in the low-cycle fatigue region to 30.4 million in the high-cycle fatigue region. The UTS for this material was determined as 632 MPa, while the UCS was 402 MPa. More information about this material system and the testing conditions along with more data for further analyses can be found in [47].

Five of the seven existing S–N curves plus the static strengths were used as input data for all the CLDs, except for the linear and Kawai models which require only one S–N curve. The remaining two were used to evaluate the modeling accuracy of the proposed methods.

For material DD16, the strength ratio according to Kawai is −0.63. Therefore, the S–N curve at $R = -0.5$ was also used for the derivation of the CFL based on Kawai's instructions. Preprocessing of the experimental data showed that the behavior of parameter f in Harris's model can be better fitted by the power law given by Eq. 4.14. Equations 4.12 and 4.13 were used for the other two model parameters. However, application of the model based on the linear fitting of all three parameters was also performed. Power S–N curves were used for the implementation of all models, but again, Boerstra's CLD was constructed using the untreated experimental data as well.

Constant life diagrams according to the described models are presented in Fig. 4.26. The linear diagram is accurate only for the prediction of the curve at the stress ratio, $R = -0.5$ ($R^2 = 0.93$), but failed to accurately predict the curve at $R = 0.8$. The predictions of the piecewise linear diagram were better in both cases. The influence of the selection of the power or linear fitting for estimation of the parameter f in Harris's model is significant, as shown in Fig. 4.26c. The bad fitting quality of Eq. 4.11 for the derivation of the relationship between parameter f and number of loading cycles results in the inaccurate CLD presented in Fig. 4.26c. On the other hand, use of the S–N curve at $R = -0.5$, as being closest to the S–N curve determined as critical by Kawai, seems to improve the modeling accuracy, although not in a consistent manner. Furthermore, the application of Boerstra's model based on fitted S–N data significantly improved accuracy, especially for the

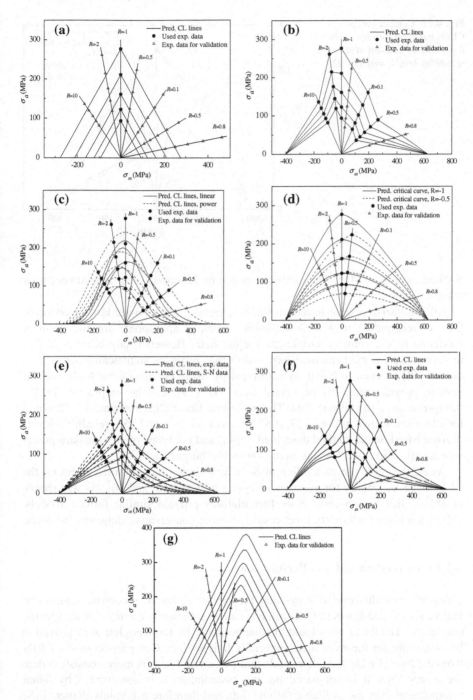

Fig. 4.26 Constant life diagrams for $N = 10^3$–10^7, material #3 (linear-a, piecewise linear-b, Harris-c, Kawai-d, Boerstra-e, piecewise non-linear-f, Kassapoglou-g)

Fig. 4.27 Comparison of CLDs based on different S–N formulations and different reliability levels, material #2

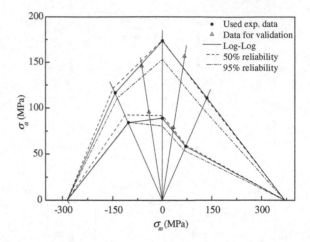

S–N curve at $R = -0.5$. The model proposed by Kassapoglou produced very poor results.

All comparisons were based on the assumption that a power law-based equation, corresponding to a 50% reliability level, is appropriate for the accurate modeling of the constant amplitude fatigue data. However, any other S–N formulation, e.g., a mathematical expression that provides statistically based S–N curves [48] or even S–N curves estimated by using computational tools such as genetic programming [8] or neural networks [19], can be employed for the interpretation of the fatigue data. The piecewise linear CLD for material #2 would look like that shown in Fig. 4.27, if S–N curves for 50% and 95% reliability levels derived based on the method described in [48] and not based on the standard power low equation are used for interpretation of the fatigue data.

As can be seen, use of a different S–N formulation has a limited effect on the CLD shape, especially for numbers of cycles between 10^3 and 10^7. As was shown in [8], in this region most S–N formulations provide similar fatigue models, although a higher reliability level results in more conservative diagrams however.

4.3.2.1 Evaluation of CLD Performance

Comparison of the results shows that the piecewise linear, piecewise non-linear, Harris, Kawai and Boerstra CLD models can be sufficiently accurate under specific conditions. The linear model and the one proposed by Kassapoglou were proved to be inaccurate for the examined material's fatigue data. Comparison of the CLDs from the four eligible models reveals that piecewise linear is more consistent than the others since it is not based on any assumption. It is constructed by linear interpolation over the available fatigue data and therefore accurately depicts their behavior. The other three diagrams are very sensitive to the selection of the input data, especially Kawai, and estimation of the model parameters, e.g., Harris.

Fig. 4.28 Predicted S–N
curves for $R = 0.5$,
material #1

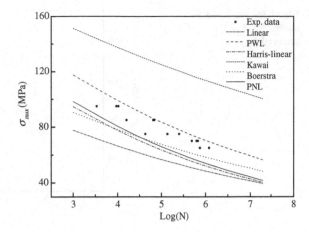

Fig. 4.29 Predicted S–N
curves for $R = -0.4$,
material #2

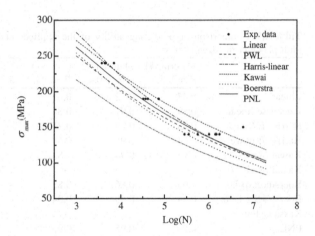

A graphical comparison of the predicting ability of the different CLD modeling methods is attempted in Figs. 4.28, 4.29, 4.30 and compared to the available experimental data for arbitrarily selected cases. The linear model underestimates the fatigue strength of the examined material in all the examined cases, leading to conservative fatigue life predictions. On the other hand, Kawai's model generally overestimates the behavior, implying an optimistic assessment of fatigue life and thus a non-conservative fatigue design. Although still not accurate, the predictions from the piecewise linear and the piecewise non-linear together with those derived by the Boerstra model seem to be the most representative of the fatigue behavior of the material under the specific loading pattern.

A quantification of the predicting ability of each of the applied models was performed. Table 4.2 shows the R^2 values between the predicted curves and the experimental data for validation.

Generally, higher values were exhibited by the piecewise non-linear followed by the piecewise linear formulation. These models seem to be the most reliable for

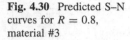

Fig. 4.30 Predicted S–N
curves for $R = 0.8$,
material #3

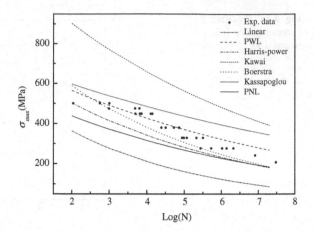

Table 4.2 Comparison of predicting ability of the applied models in terms of coefficient of multiple determination (R^2)

	Material #1	Material #2		Material #3		
	$R = 0.5$	$R = -0.4$	$R = -2.5$	$R = -1$	$R = -0.5$	$R = 0.8$
Linear	0.37	0.72	0.61	–	0.93	0.35
Piecewise linear	0.89	0.91	0.84	–	0.88	0.93
Harris–f: Linear	0.63	0.95	0.71	–	0.77	0.51
Harris–f: power law	0.64	0.94	0.64	–	0.94	0.80
Kawai ($R = -1$)	0.15	0.94	0.83	–	0.87	0.31
Kawai ($R = -0.5$)	–	–	–	0.76	–	0.43
Boerstra-Exp Data	0.65	0.86	0.83	–	0.60	0.85
Boerstra-SN Data	0.69	0.89	0.84	–	0.84	0.91
Kassapoglou	–	–	–	0.93	0.41	0.48
PNL	0.88	0.95	0.96		0.92	0.78

the entire set of data examined in the present chapter. Other CLD formulations can also be accurate, however, although not consistently so. Their accuracy depends on the quality of the examined fatigue data, the selected input data (e.g., linear, Kawai) and the quality of the fitting and/or optimization for estimation of the parameters (Harris and Boerstra).

In terms of need for experimental data, it is obvious that the model proposed by Kassapoglou is the least demanding, followed by the linear and Kawai models. However, as already discussed, this compromise reduces the accuracy of the predictions. The piecewise linear can also be implemented by using a single S–N curve, thus becoming equivalent to a shifted Goodman diagram, but, as mentioned above, in this case its predictive ability is also reduced.

As far as ease of application is concerned, the only difficulty occurs in the Harris and Boerstra models. According to the former, a non-linear regression should be performed for the derivation of the three-model parameters, while for the latter, a five-parameter optimization problem must be solved to estimate desired constant lifelines.

Apart from the piecewise linear, all other models depend on a number of assumptions. These assumptions originate either from experience and experimental evidence, e.g., in the linear, Harris, Kawai and Boerstra models, or are clearly theoretical assumptions like the one introduced by Kassapoglou. As previously shown, the adoption of any assumption can simplify the models, which, under certain conditions, can produce quite accurate results.

However, there is no guarantee that these models can be used for different materials or even different loading patterns.

4.3.3 Concluding Remarks Regarding the CLD Performance

A comparison of the commonly used and recently developed models for the derivation of constant life diagrams for composite materials was carried out in this section. Seven methods were described, and their prediction accuracy was evaluated over a wide range of constant amplitude fatigue data obtained from GFRP materials. The influence of the selection of the CLD method on the fatigue life prediction of composite materials was quantified. The following conclusions were drawn:

- The selection of an accurate CLD formulation is essential for the overall accuracy of a fatigue life prediction methodology. As shown, the "wrong" choice can produce very conservative or very optimistic S–N curves, which is directly reflected in the corresponding life assessment.
- All methods involve the problem of mixing static and fatigue data. Their accuracy is reduced when curves close to $R = 1$ (in tension or compression) have to be predicted. Moreover, the same applies for the derivation of accurate S–N curves to describe the very low-cycle fatigue regime, i.e., $N < 100$. The unified equation used in Harris's model to describe fatigue behavior for both tension and compression loading also includes the influence of the damage mechanisms that developed under different loading patterns. All other models work separately for tension and compression loading; they are based on different equations for the description of different parts of the constant life diagram.
- The simplicity offered by some of the models, e.g., linear, Kassapoglou, Kawai, in most cases compromises accuracy. In addition, although these models were developed with the aim of minimizing the amount of experimental data required, they do not offer the possibility of using more data when an extensive database is available, e.g., linear and Kawai are based on the critical S–N curve and cannot accommodate any other S–N curves in order to improve the accuracy of the predictions. Another deficiency of models like these two is that they cannot be used to analyze random variable amplitude fatigue loading with continuously changing mean and amplitude, since they are accurate only

for S–N curves close to the critical one (for the linear) and only if the critical one (according to the material) can be experimentally derived (Kawai). It should be mentioned however, that Kawai introduced his model for the description of CFRP material behavior different from that of the GFRP materials examined in this study.

- The accuracy of Harris's model is acceptable only when the behavior of the model parameters can be effectively fitted versus the fatigue life. However, the fitting methods proposed by Harris do not always lead to accurate results.
- Boerstra's model can be used without the need to fit the available experimental data with an S–N curve. This is an asset since it can therefore be used to model variable amplitude data with continuously varying mean and amplitude values. However, it was proved that use of the fitted S–N data instead improves the modeling accuracy of the Boerstra model.
- The relatively simple piecewise linear and its non-linear counterpart, the PNL, were proved the most accurate of the compared formulations when a reasonable number of S–N curves (≥2–3) is available. A more sophisticated, non-linear interpolation between the known S–N values and a more realistic description of the behavior close to $R = 1$ would improve the results of these models.

4.4 Stiffness Degradation

The stiffness/strength-based models were mostly established as phenomenological models since they propose an evolutionary law to describe the gradual degradation of the specimen's stiffness or strength in terms of macroscopically measurable properties [49]. The modeling of the varying damage metric reflects the damage accumulation in specimens during fatigue. The damage metric depends on many factors, including applied cyclic stress, number of fatigue cycles, loading frequency and environmental conditions.

The stiffness-based model is derived from the change in stiffness of a material or a structural component undergoing fatigue. The residual stiffness is expressed as a function of initial stiffness and number of cycles. The relationship between these three parameters can be of any mathematical form, e.g., linear, power, sigmoid, depending on the experimental data. Similar models were developed using residual strength as the damage metric. However, stiffness offers certain advantages compared to strength: it can be measured using non-destructive methods and presents less scatter on the measured results than strength data. Furthermore, residual strength exhibits only minimal decreases with the number of cycles until it begins to change rapidly close to the end of lifetime, while stiffness exhibits greater changes during fatigue life [50–53] and thus a higher sensitivity to damage progression.

A wide variety of composite materials exhibit a stiffness degradation trend that can be simulated by a curve like the one schematically shown in Fig. 4.31.

Fig. 4.31 Schematic representation of a typical stiffness degradation curve for composite materials

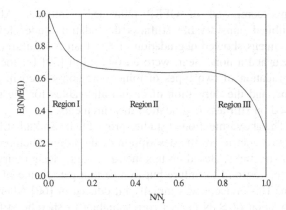

The three regions designated in this figure were firstly distinguished by Schulte for the tension-tension fatigue of cross-ply carbon/epoxy laminates [54].

In the initial region, and up to around 10% of fatigue life, the material exhibits a sudden stiffness reduction (compared to region II). In the intermediate region, the material's stiffness degrades at a constant and moderate rate. Finally, significant deterioration of the material can be observed close to the end of the fatigue life. A third region with a steeply descending slope simulates this phenomenon.

The main objective of the research community is to model this behavior for any selected composite material for different structural applications. To this end, a power law relationship was used in [55, 56] to describe the stiffness degradation of a glass-fiber cloth composite laminate and the concept of fatigue modulus was introduced. It was defined as the ratio of maximum stress over maximum strain at a specific cycle. The fatigue modulus measured during the first loading cycle was assumed to be the same as the elastic modulus. The fatigue modulus at failure was dependent on the applied cyclic stress level.

A modified exponential model was introduced in [3] to describe the behavior of graphite/epoxy laminates. Stiffness at any loading cycle was expressed as a function of the initial stiffness, ultimate strength of the laminate, applied stress level and two constants that should be adjusted by fitting the model to the experimental data.

Although a number of studies have been presented on the modeling of the stiffness degradation of several materials, fewer have been presented on the investigation of the fatigue behavior of structural components such as structural joints, e.g., [57, 58].

The concept of the shear stiffness modulus was introduced in [57] for the study of the fatigue life of adhesive lap joints produced from bi-directional woven E-glass fibers and polypropylene matrix. The shear stiffness modulus was defined as the ratio between shear stress in the bond and axial strain. Experimental results showed that the joints exhibited little and almost linear stiffness degradation throughout most of their life, followed by a sudden decrease between $0.95 < N/N_f < 1$. A similar tendency was observed in [58] for double-lap joints composed of GFRP pultruded laminates. Although GFRP laminates exhibited considerable stiffness degradation under low cyclic loads, the joint specimens showed degradation of less than 5% failure. Two empirical models, a linear and a non-linear, were introduced in [59] for the modeling of the stiffness degradation of two types of joints commonly used in civil engineering applications, and the derivation of design allowables for the examined structural joints based on stiffness degradation measurements.

The above-mentioned studies proved that residual stiffness could be an efficient damage metric for the description of the fatigue behavior of composite materials and structures. Based on this metric, fatigue design curves can be derived that do not correspond to failure but to a certain percentage of specimen stiffness reduction. This concept was initially introduced in [60] where the author proposed the derivation of S–N curves corresponding to specific stiffness degradation. In this case, data points in the S–N plane denote that under cyclic stress a predetermined stiffness reduction is reached after N cycles. Such stiffness-controlled fatigue design curves, henceforth denoted by Sc-N, can be derived in a straightforward manner using empirical stiffness degradation models, like the simple one previously introduced in [61]. The accuracy of this theoretical approach has been validated independently using experimental data from different material systems [62].

Especially for the design of structures containing rotating parts, like wind turbine or helicopter rotor blades, Sc-N curves can better serve the requirements of proposed full-scale testing procedures [63], where blade functional failure is said to correspond to irreversible stiffness reduction of up to 10%. Therefore, to conform to this kind of testing procedure for example, fatigue design allowables in the form of Sc-N curves must be established, and to that end systematic stiffness reduction data monitoring and statistical analysis must be performed beforehand.

In the following, the experimental data presented in Chap. 2 will be used for the demonstration of a method for the derivation of Sc-N curves, and the modeling of the fatigue life of the examined material. Sc-N curves determined for each R-value and off-axis direction will be compared to fatigue strength ones-N curves derived based on the statistical analysis of the fatigue strength data.

4.4.1 Fatigue Life Modeling Based on Stiffness Degradation

Although more complicated and therefore more accurate models for the modeling of stiffness variations during fatigue life exist [54], the simple empirical model for the description of stiffness changes and the derivation of stiffness-controlled design curves previously introduced in [61] and further validated for different material systems in [62] is used here for the demonstration of the technique. A brief outline of the model is given below.

The degree of damage in a polymer matrix composite coupon can be evaluated by measuring stiffness degradation, $E(N)/E(1)$, where $E(1)$ denotes the Young's modulus of the material measured at the first cycle, different in general from the static value, E_{st}, and $E(N)$ is the Young's modulus measured at the N-th cycle. It is assumed that stiffness degradation can be expressed by [53]:

$$\frac{E(N)}{E(1)} = 1 - k_1 \left(\frac{\sigma_a}{E_{st}}\right)^{k_2} N \qquad (4.34)$$

Material constants, k_1 and k_2, in Eq. 4.34 are determined by curve fitting of the respective experimental data for $E(N)/E(1)$, which depend on the number of stress cycles, N, and the level of applied cyclic stress amplitude, σ_a. Rearranging Eq. 4.34 in the following form:

$$\frac{1 - \frac{E(N)}{E(1)}}{N} = k_1 \left(\frac{\sigma_a}{E_{st}}\right)^{k_2} \qquad (4.35)$$

allows the easy determination of model constants.

Equation (4.34) also establishes a stiffness-based design criterion since for a predetermined value of $E(N)/E(1)$, e.g., p, one can solve for σ_a to obtain an alternative form of design curve, Sc-N, corresponding not to material failure but to a specific stiffness degradation percentage $(1 - p)\%$.

Sc-N curves for any specific stiffness degradation level, $E(N)/E(1)$, can be easily calculated by means of the following equation:

$$\sigma_a = E_{st} \left(\frac{1 - \frac{E(N)}{E(1)}}{k_1 N}\right)^{\frac{1}{k_2}} \qquad (4.36)$$

4.4.2 Stiffness-Based and Reliability S–N Curves

Based on stiffness degradation data (Chap. 2), stiffness-controlled Sc-N curves, corresponding to specific $E(N)/E(1)$ values, were calculated by means of Eq. 4.36. Fatigue strength curves were also defined at predetermined survival probability values using Eq. 4.4 and parameters of the statistical model shown in Table 4.1.

When comparing these two kinds of fatigue design curves it is concluded that there is a correlation between the probability level of the fatigue strength curves and the stiffness degradation level of the Sc-N curves. To any survival probability level, $P_S(N)$, there is a corresponding unique stiffness degradation value, $E(N)/E(1)$, which can be determined from the cumulative distribution function of the respective stiffness degradation data. It is this value of $E(N)/E(1)$ for which the cumulative distribution function, $F(E(N)/E(1))$, takes the value of $P_S(N)$. For example, $F(E(N)/E(1)) = 0.95$ for a residual stiffness of 0.96 for 15° off-axis specimens loaded under $R = 0.1$, as shown in Fig. 4.28, and therefore, the Sc-N curve for $E(N)/E(1) = 0.96$ is corroborated well by the fatigue strength curve

Fig. 4.32 Sampling distribution of stiffness degradation data, $R = 0.1$, 15° off-axis

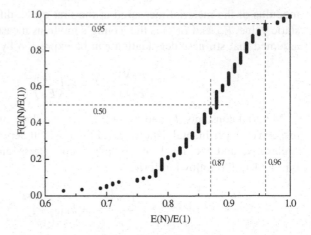

Fig. 4.33 Sc-N vs. S–N curves. $R = 0.1$, 45° off-axis

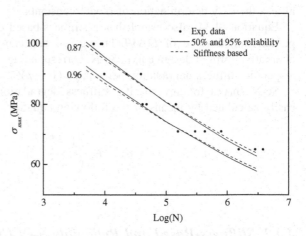

corresponding to a 95% reliability level as shown in Fig. 4.29. The same applies for the reliability level of 50%, a residual stiffness of 0.87 corresponds to this value of cumulative distribution function (see Fig. 4.32). It is indeed observed that Sc-N and S–N curves from each set lie very close to each other and that the former type of design curve is slightly more conservative. Using the Sc-N at $E(N)/E(1) = 0.96$ as derived from Fig. 4.32 as design allowable, a reliability level of at least 95% is guaranteed while stiffness reduction will be less than 5%. Similar comments are also valid for Figs. 4.33, 4.34 where corresponding curves are shown for specimens cut at different off-axis angles and tested under different R-ratios.

Observing the two different curves derived as stated in the above, it was concluded that they are similar for all cases considered in this work, with the Sc-N being slightly more conservative in general. Therefore, in design an Sc-N curve providing information on both survival probability and residual stiffness can be used.

It should be mentioned that this good correlation between stiffness-based and reliability S–N curves applies to all other types of specimens, tested under different

Fig. 4.34 Sc-N vs. S–N curves. $R = 10$, 90° off-axis

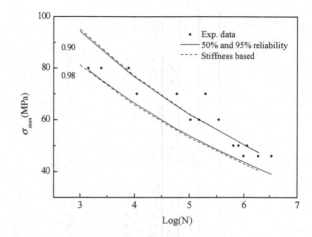

loading conditions. In Table 4.3, S–N curve equations are given for a 95% reliability level (according to Whitney's method, see Chap. 3) for all datasets used in this study and are compared to the corresponding stiffness-based Sc-N curve equations.

Despite the observed discrepancies, which are not significant in most cases, stiffness-based Sc-N curves can be used instead of reliability S–N curves in design. Curves of the former type provide information regarding two design parameters, reliability and stiffness degradation level. Thus, they can be used in design to fulfill the requirements of design codes and regulations. In addition, Sc-N curves can be determined much faster, as stiffness degradation trends are readily captured by testing only a small number of specimens.

To demonstrate this, the procedure for the determination of stiffness-based Sc-N curves was repeated by using only half of the specimens. Half of the specimens from each set was randomly selected and the calculations were repeated. The Sc-N curves determined in this way were then compared to the original ones. The probability cumulative distributions were almost identical in most of the cases studied, e.g., see Fig. 4.32. Thus, the Sc-N curves were similar to those determined by using the full dataset as shown for example in Fig. 4.35 for 30° off-axis coupons, tested under alternating stress, $R = -1$. (Figures 4.36, 4.37).

4.4.3 Concluding Remarks

During operational life, the stiffness of a structural element is reduced. The modeling of the fatigue life of specimens cut at several off-axis angles from a multidirectional laminate $[0/(\pm 45)_2/0]_T$ and subjected to uniaxial cyclic loading over a wide range of R-ratios was performed in this chapter based on stiffness degradation measurements.

A simple empirical model was used for the determination of design curves, which do not correspond to fatigue strength but to a predetermined value of

Table 4.3 Stiffness-controlled and S–N curves for 95% survival probability*

Direction	R-ratio							
	10		−1		0.1		0.5	
	Sc-N	S-N	Sc-N	S-N	Sc-N	S-N	Sc-N	S-N
0°	$305.1\,N^{-0.0591}$ (0.97)	$244.51\,N^{-0.0387}$	$164.6\,N^{-0.0585}$ (0.98)	$142.2\,N^{-0.0463}$	$585.4\,N^{-0.1116}$ (0.85)	$528.5\,N^{-0.1008}$	$366.4\,N^{-0.0502}$ (0.95)	$326.4\,N^{-0.0420}$
15°					$168.2\,N^{-0.0733}$ (0.96)	$164.9\,N^{-0.0694}$		
30°	$327.7\,N^{-0.1139}$ (0.99)	$321.9\,N^{-0.1115}$	$113.2\,N^{-0.0807}$ (0.96)	$113.8\,N^{-0.0788}$				
45°	$216.9\,N^{-0.0758}$ (0.98)	$238.8\,N^{-0.0843}$	$133.4\,N^{-0.0850}$ (0.95)	$112.9\,N^{-0.0721}$	$161.6\,N^{-0.0950}$ (0.98)	$153.4\,N^{-0.0921}$	$166.9\,N^{-0.0719}$ (0.93)	$156.96\,N^{-0.0671}$
60°	$154.2\,N^{-0.0953}$ (0.98)	$143.5\,N^{-0.0833}$	$121.7\,N^{-0.0986}$ (0.89)	$121.8\,N^{-0.0993}$				
75°					$94.47\,N^{0.0995}$ (0.83)	$75.03\,N^{-0.0769}$		
90°	$67.20\,N^{-0.0441}$ (0.99)	$70.39\,N^{-0.0485}$		$77.46\,N^{-0.0712}$	$70.62\,N^{-0.0963}$ (0.84)	$55.41\,N^{-0.0690}$		

* Numbers in parentheses indicate the respective $E(N)/E(1)$ values

Fig. 4.35 Sc-N vs. S–N curves. $R = 0.1$, 15° off-axis

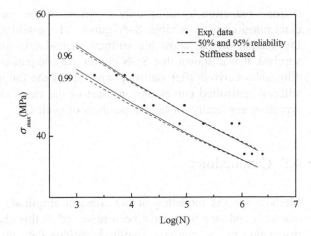

Fig. 4.36 Sampling distributions of complete and half of dataset. $R = -1$, 30° off-axis

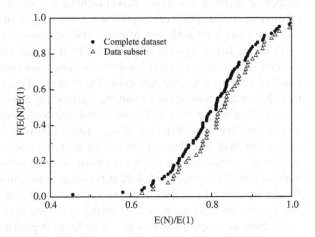

Fig. 4.37 Comparison of Sc-N curves determined using all and half of experimental stiffness degradation data

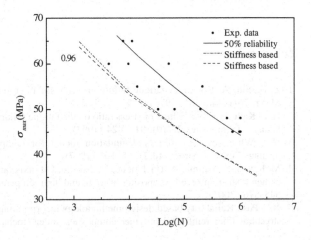

stiffness reduction by using only a portion of the fatigue data required for the determination of a reliable S–N curve. The established Sc-N curves provide information on the allowable stiffness degradation and also the probability of survival. It was shown that Sc-N curves are comparable to corresponding design allowables derived after statistical analysis of the fatigue strength data, although stiffness-controlled curves are, in most of the cases studied, slightly more conservative, especially for higher numbers of cycles.

4.5 Conclusions

Methods for the modeling of the constant amplitude fatigue life of composite materials and structures have been reviewed in this chapter. The traditional representation of the constant amplitude fatigue data on the S–N plane has been addressed in the first sections. Novel techniques that can be used for the derivation of more accurate S–N curve types were introduced and their modeling accuracy has been compared with that of conventional ways of representing fatigue data.

Constant life diagrams are commonly used for the prediction of "unseen" material data, under different loading conditions from those for which experimental constant amplitude fatigue data exist. The concept of the CLDs has been presented in this chapter and conclusions about their predicting ability have been drawn.

Modeling of the fatigue life of fiber-reinforced composite materials based on non-destructive measurements of the material stiffness and its fluctuations during fatigue life has long been a subject of investigation. A simple model has been employed in this chapter for the demonstration of a method that allows the derivation of stiffness-controlled S–N curves that represent the material's constant life fatigue behavior by accommodating both strength and stiffness information. The use of this method enables the testing time required to obtain fatigue design allowables, corresponding to a preset stiffness degradation and reliability level, to be reduced by at least 50%.

References

1. Z. Hashin, A. Rotem, A fatigue failure criterion for fiber–reinforced materials. J. Compos. Mater. **7**(4), 448–464 (1973)
2. H. El Kadi, F. Ellyin, Effect of stress ratio on the fatigue failure of fiberglass reinforced epoxy laminae. Composites **25**(10), 917–924 (1994)
3. H.A. Whitworth, A stiffness degradation model for composite laminates under fatigue loading. Compos. Struct. **40**(2), 95–101 (1997)
4. M. Kawai, S. Yajima, A. Hachinohe, Y. Takano, Off–axis fatigue behavior of unidirectional carbon fiber–reinforced composites at room and high temperatures. J. Compos. Mater. **35**(7), 545–576 (2001)
5. N.L. Post, Reliability based design methodology incorporating residual strength prediction of structural fiber reinforced polymer composites under stochastic variable amplitude fatigue

loading, PhD Thesis, Virginia Polytechnic Institute and State University, March 18, Blacksburg, Virginia (2008)

6. A.P. Vassilopoulos, R. Bedi, Adaptive neuro-fuzzy inference system in modeling fatigue life of multidirectional composite laminates. Comp. Mater. Sci. **43**(4), 1086–1093 (2008)

7. A.P. Vassilopoulos, E.F. Georgopoulos, T. Keller, Genetic programming in modeling of fatigue life of composite materials. in *13th International Conference on Experimental Mechanics-ICEM13: Experimental Analysis of Nano and Engineering Materials and Structures, Alexandroupolis*, Greece, July 1–6, 2007

8. A.P. Vassilopoulos, E.F. Georgopoulos, T. Keller, Comparison of genetic programming with conventional methods for fatigue life modeling of FRP composite materials. Int. J. Fatigue **30**(9), 1634–1645 (2008)

9. R.P.L. Nijssen, O Krause, T.P Philippidis, Benchmark of lifetime prediction methodologies, Optimat blades technical report, 2004, OB_TG1_R012 rev.001, http://www.wmc.eu/public_docs/10218_001.pdf

10. T. Adam, G. Fernando, R.F. Dickson, H. Reiter, B. Harris, Fatigue life prediction for hybrid composites. Fatigue **11**(4), 233–237 (1989)

11. J.A. Epaarachchi, P.D. Clausen, An empirical model for fatigue behavior prediction of glass fiber reinforced plastic composites for various stress ratios and test frequencies. Compos. Part A-Appl. Sci. **34**(4), 313–326 (2003)

12. J.M. Whitney Fatigue characterization of composite materials. in *Fatigue of Fibrous Composite Materials*, ASTM STP 723, American Society for Testing and Materials, 1981, 133–151

13. G.P. Sendeckyj, Fitting models to composite materials, in *Test methods and design allowables for fibrous composites, (ASTM STP 734*, ed. by C.C. Chamis (American Society for Testing and Materials, West Conshohocken, PA, 1981), pp. 245–260

14. J.A. Lee, D.P. Almond, B. Harris, The use of neural networks for the prediction of fatigue lives of composite materials. Compos. Part A-Appl. Sci. **30**(10), 1159–1169 (1999)

15. Y. Al-Assaf, H. El Kadi, Fatigue life prediction of unidirectional glass fiber/epoxy composite laminae using neural networks. Compos. Struct. **53**(1), 65–71 (2001)

16. J.A. Lee, D.P. Almond, A neural-network approach to fatigue-life prediction, in *Fatigue in composites*, ed. by B. Harris (Woodhead Publishing Ltd, Cambridge, UK, 2003), pp. 569–589

17. A.P. Vassilopoulos, E.F. Georgopoulos, V. Dionysopoulos, Modeling fatigue life of multidirectional GFRP laminates under constant amplitude loading with artificial neural networks. Adv. Compos. Lett. **15**(2), 43–51 (2006)

18. R.C.S.F. Junior, A.D.D. Neto, E.M.F. Aquino, Building of constant life diagrams of fatigue using artificial neural networks. Int. J. Fatigue **27**(7), 746–751 (2005)

19. A.P. Vassilopoulos, E.F. Georgopoulos, V. Dionysopoulos, Artificial neural networks in spectrum fatigue life prediction of composite materials. Int. J. Fatigue **29**(1), 20–29 (2007)

20. C.S. Lee, W. Hwang, H.C. Park, K.S. Han, Failure of carbon/epoxy composite tubes under combined axial and torsional loading 1. Experimental results and prediction of biaxial strength by the use of neural networks. Compos. Sci. Technol. **59**(12), 1779–1788 (1999)

21. J. Jia, J.G. Davalos, An artificial neural network for the fatigue study of bonded FRP-wood interfaces. Compos. Struct. **74**(1), 106–114 (2006)

22. M.A. Jarrah, Y. Al-Assaf, H. El Kadi, Neuro-Fuzzy modeling of fatigue life prediction of unidirectional glass fiber/epoxy composite laminates. J. Compos. Mater. **36**(6), 685–699 (2002)

23. AIM Learning Technology, http://www.aimlearning.com, last update 10.01.2007

24. J.R. Koza, *Genetic Programming on the Programming of Computers by Means of Natural Selection* (MIT Press, Cambridge, MA, 1992)

25. J.R. Koza, Genetic Programming, in *Encyclopaedia of Computer Science and Technology*, ed. by J.G. Williams, A. Kent (Marcel-Dekker, NY, 1998), pp. 29–43. 39. Supplement 24

26. V. Babovic, M. Keijzer, Genetic programming as a model induction engine. J. Hydroinform **2**(1), 35–60 (2000)

27. A.P. Vassilopoulos, B.D. Manshadi, T. Keller, Influence of the constant life diagram formulation on the fatigue life prediction of composite materials. Int. J. Fatigue **32**(4), 659–669 (2009)

28. N. Gathercole, H. Reiter, T. Adam, B. Harris, Life prediction for fatigue of T800/5245 carbon fiber composites: I. Constant amplitude loading. Int. J. Fatigue **16**(8), 523–532 (1994)
29. M.H. Beheshty, B. Harris, A constant life model of fatigue behavior for carbon fiber composites: the effect of impact damage. Compos. Sci. Technol. **58**(1), 9–18 (1998)
30. M. Kawai, M. Koizumi, Nonlinear constant fatigue life diagrams for carbon/epoxy laminates at room temperature. Compos.: Part A **38**(11), 2342–2353 (2007)
31. J.F. Mandell, D.D. Samborsky, L. Wang, N.K. Wahl, New fatigue data for wind turbine blade materials. J. Sol. Energy Eng. Trans. ASME **125**(4), 506–514 (2003)
32. H.J. Sutherland, J.F. Mandell, Optimized constant life diagram for the analysis of fiberglass composites used in wind turbine blades. J. Sol. Energy Eng. Trans. ASME **127**(4), 563–569 (2005)
33. T.P. Philippidis, A.P. Vassilopoulos, Complex stress state effect on fatigue life of GFRP laminates Part I, Experimental. Int. J. Fatigue **24**(8), 813–823 (2002)
34. B. Harris, A parametric constant-life model for prediction of the fatigue lives of fiber-reinforced plastics, in *Fatigue in Composites*, ed. by B. Harris (Woodhead Publishing Limited, Cambridge, UK, 2003), pp. 546–568
35. M. Kawai, A method for identifying asymmetric dissimilar constant fatigue life diagrams for CFRP laminates. Key. Eng. mater. **334–335**, 61–64 (2007)
36. G.K. Boerstra, The multislope model: a new description for the fatigue strength of glass reinforced plastic. Int. J. Fatigue **29**, 1571–1576 (2007)
37. C. Kassapoglou, Fatigue life prediction of composite structures under constant amplitude loading. J. Compos. Mater. **41**(22), 2737–2754 (2007)
38. C. Kassapoglou, Fatigue of composite materials under spectrum loading. Compos. Part A-Appl. Sci. **41**(5), 663–669 (2010)
39. A.P. Vassilopoulos, B.D. Manshadi, T. Keller, Piecewise non-linear constant life diagram formulation for FRP composite materials. Int. J. Fatigue **32**(10), 1731–1738 (2010)
40. W.Z. Gerber, Bestimmung der zulässigen spannungen in eisen-constructionen (Calculation of the allowable stresses in iron structures). Z Bayer Archit. Ing-Ver **6**(6), 101–110 (1874)
41. J. Goodman, *Mechanics Applied to Engineering* (Longman Green, Harlow, 1899)
42. V.A. Passipoularidis, T.P. Philippidis, A study of factors affecting life prediction of composites under spectrum loading. Int. J. Fatigue **31**, 408–417 (2009)
43. T.P. Philippidis, A.P. Vassilopoulos, Life prediction methodology for GFRP laminates under spectrum loading. Compos Part A-Appl S **35**(6), 657–666 (2004)
44. Awerbuch J, Hahn HT. Off-axis fatigue of graphite/epoxy composite. in *Fatigue of Fibrous Composite Materials. ASTM STP 723*, (American Society for Testing and Materials, 1981), pp. 243–273
45. M. Kawai, T. Murata, A three-segment anisomorphic constant life diagram for the fatigue of symmetric angle-ply carbon/epoxy laminates at room temperature. Compos Part A-Appl S **41**(10), 1498–1510 (2010)
46. Nijssen RPL. OptiDAT–fatigue of wind turbine materials database, 2006. <http://www.kc-wmc.nl/optimat_blades/index.htm>
47. Mandell JF, Samborsky DD. DOE/MSU Composite Material Fatigue Database. Sandia National Laboratories, SAND97-3002 (online via www.sandia.gov/wind, v. 18, 21st March 2008 Updated)
48. G.P. Sendeckyj. Fitting models to composite materials fatigue data. Test Methods and Design Allowables for Fibrous Composites. ASTM STP 734. C.C. CHAMIS, editor. American Society for Testing and Materials, 1981. p. 245–260
49. J. Degrieck, W.M. Paepegem, Fatigue damage modeling of fiber-reinforced composite materials: a review. Appl. Mech. Rev. **54**(4), 279–300 (2001)
50. G.P. Sendeckyj. Life prediction for resin-matrix composite materials. in *Fatigue of Composite Materials*, ed. by K.L. Reifsneider, Composite Materials Series 4 (Elsevier, Amsterdam, 1991)
51. A.L. Highsmith, K.L. Reifsneider, Stiffness Reduction Mechanisms, in *Composite Laminates, Damage in Composite Materials, ASTM STP 775*, ed. by K.L. Reifsneider (American Society for Testing and Materials, West Conshohocken, PA, 1982), pp. 103–117

52. R. Talreja, *Fatigue of Composite Materials* (Technomic, Lancaster Pennsylvania, 1987)
53. S.I. Andersen, P. Brondsted, H. Lilholt, Fatigue of polymeric composites for wingblades and the establishment of stiffness-controlled fatigue diagrams. in *Proceedings of 1996 European Union Wind Energy Conference*, Göteborg, Sweden, (20–24 May 1996) pp. 950–953
54. W. Van Paepegem, Fatigue damage modeling of composite materials with the phenomenological residual stiffness approach, in *Fatigue Life Prediction of Composites and Composite Structures*, ed. by A.P. Vassilopoulos (Woodhead Publishing Ltd., Cambridge, 2010)
55. W. Hwang, K.S. Han, Fatigue of composites-fatigue modulus concept and life prediction. J. Compos. Mater. **20**(2), 154–165 (1986)
56. H.T. Hahn, R.Y. Kim, Fatigue behavior of composite laminate. J. Compos. Mater. **10**(2), 156–180 (1976)
57. J.A.M. Ferreira, P.N. Reis, J.D.M. Costa, M.O.W. Richardson, Fatigue behavior of composite adhesive lap joints. Compos. Sci. Technol. **62**(10–11), 1373–1379 (2002)
58. T. Keller, A. Zhou, Fatigue behavior of adhesively bonded joints composed of pultruded GFRP adherends for civil infrastructure applications. Compos Part A-Appl S **37**(8), 1119–1130 (2006)
59. Y. Zhang, A.P. Vassilopoulos, T. Keller, Stiffness degradation and fatigue life prediction of adhesively-bonded joints for fiber-reinforced polymer composites. Int. J. Fatigue **30**(10–11), 1813–1820 (2008)
60. M.J. Salkind, Fatigue of composites, in *Composite Materials, Testing And Design, ASTM STP 497*, ed. by H.T. Corten (American Society for Testing and Materials, West Conshohocken, PA, 1972), pp. 143–169
61. T.P. Philippidis, A.P. Vassilopoulos, Fatigue design allowables for GFRP laminates based on stiffness degradation measurements. Compos. Sci. Technol. **60**(15), 2819–2828 (2000)
62. T.P. Philippidis, A.P. Vassilopoulos, Fatigue of composite laminates under off-axis loading. Int. J. Fat. **21**, 253–262 (1999)
63. Anon. IEC-TC88-WG8 test guideline: "Full-scale structural testing of rotor blades for WTGS's", IEC 61400-23 (1998)

Chapter 5
Fatigue of Adhesively-Bonded GFRP Structural Joints

5.1 Introduction

Joints are the most likely locations for failure initiation and propagation in engineering structures, since their function is to transfer loads from one part of the structure to another. The scientific community has focussed its attention on adhesively-bonded and bolted joints since they are the main types used in the composites. The fatigue of adhesively-bonded joints was initially investigated for aerospace applications since they offer many advantages compared to riveting and bolting, such as considerable mass saving, cheaper fabrication, improved aerodynamics and the avoidance of holes that act as areas of stress concentration [1] and potential moisture ingress along cut fibers. Although bonded joints are already used even as primary structural connections in aerospace applications, their potential is yet to be fully exploited in civil and other engineering domains. The main reason for this is the lack of confidence in the ability of bonded joints to act as structural connections. This skepticism is even more pronounced when the structure has to operate under fatigue loads and is subjected to aggressive environments such as high temperature and humidity. In these cases, the lack of knowledge concerning adhesive joints leads to a lack of confidence in their ability to retain their structural integrity.

During the loading of a structural joint, a crack or cracks initiate naturally and propagate along the least resistant path inside the component. Therefore, the development of damage in a joint is very much dependent on the constituent materials. The presence of aggressive environments, low or high temperatures and high humidity levels also affects the fatigue behavior of structural joints. Temperature and humidity can influence the mechanical properties of the adhesive and the matrix material of the composite adherends, possibly weakening the interface between fibers and matrix in the presence of humidity [2]. These synergistic effects of different loading factors result in complex fracture surfaces and make the analysis of the failure of composite joints a very challenging task.

A. P. Vassilopoulos and T. Keller, *Fatigue of Fiber-reinforced Composites*,
Engineering Materials and Processes, DOI: 10.1007/978-1-84996-181-3_5,
© Springer-Verlag London Limited 2011

The effect of the environment on the behavior of adhesively-bonded joints has formed the subject of several investigations in the past, e.g., [1–4], although most studies concern joints used specifically in the aerospace and/or automotive engineering domains. Simulated aircraft structural joints comprising CFRP adherends and an epoxy adhesive were fatigued under five different combinations of temperature and humidity and the results were reported in [1]. These showed that the fatigue behavior of the joints was considerably affected by environmental conditions. In addition, different failure modes were observed under different conditions. Cohesive failure of the adhesive occurred under hot, humid conditions. The failure mode changed to substrate failure in the case of ambient temperature, while very rapid propagation was observed when testing at −50°C. The authors attributed this behavior to the increased rigidity of the adhesive as it cooled. The static and fatigue behaviors of different joint types used in the aerospace industry were investigated in [2] (composite joints with film or paste adhesives, composite-to-metal joints). The same group investigated the temperature–dependent fatigue behavior of CFRP/epoxy double–lap joints over a wide temperature range of −50 to 90°C [3]. Unidirectional (UD) and multidirectional (MD) adherends were used. The MD joints were shown to be stronger at low temperatures, at which, according to the authors, joint strength was determined by the peak stresses. UD joints on the other hand were stronger at high temperatures where the strength was governed by the creep of the joints, determined by the minimum developed stresses. The fatigue damage and failure mechanism of single-lap joints composed of E-glass/polyethylene adherends and an ethyl-cyanoacrylate adhesive were investigated in [4]. The specimens were preconditioned for up to 90 days in water at different temperatures prior to testing. A significant reduction in fatigue strength was observed with increased immersion time and when the water temperature exceeded the glass transition temperature of the adhesive this reduction was accelerated.

Although much research effort has been devoted to the characterization of the fatigue behavior of adhesively-bonded composite joints and composite laminates under different temperature and humidity environments, there is no common method in the literature for the modeling and/or prediction of such behavior. A few modeling approaches have been published, e.g., [5–7], but, in order to accommodate a significant number of parameters that affect the fatigue life of FRP joints, these phenomenological models adopt many assumptions, e.g., [7]. Therefore, their applicability could not be validated on different material system data.

For this purpose, new computational methods have been presented in the literature [8–11] and it has been proved that they can be used to accurately model the fatigue life of composite laminates and adhesively-bonded joints [12] under various loading patterns.

In the present chapter, the fatigue behavior of double-lap joints, comprising pultruded GFRP laminates bonded by a paste epoxy adhesive system, is investigated. Axial cyclic tests were performed under four environmental conditions and the effects of low and high temperatures and high humidity evaluated. The stiffness degradation of the specimens was linked to the fatigue life via a linear stiffness degradation criterion. The dependence of stiffness changes on applied stress level

Fig. 5.1 Geometric
configuration of test
specimen

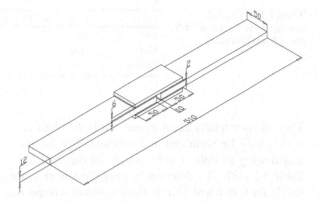

was systematically investigated and the damage accumulation in these structural elements was monitored during fatigue life. Design allowables based on stiffness degradation were established. The genetic programming tool, as described in the previous chapter, is used here to model the fatigue life of adhesively-bonded FRP joints subjected to tensile loading under different environmental conditions.

5.2 Experimental Program

Balanced adhesively-bonded double-lap joints (DLJs), composed of pultruded GFRP laminates bonded by an epoxy adhesive system, were tested under axial tensile fatigue loads in four different environments. The objective of the experimental program was to demonstrate the influence of temperature and humidity on the fatigue behavior of the examined structural components. For all the examined cases, the fatigue life corresponding to the applied load was experimentally derived. The fatigue data will be presented in the following as number of cycles to failure vs. maximum cyclic nominal section stresses (load over laminate cross-section) for the representation of the S–N curves.

The geometry of the examined joint configuration is shown schematically in Fig. 5.1. The laminate widths were 50 mm and the thicknesses were 12 and 6 mm. The overlap length was 50 mm and the adhesive thickness 2 mm. A relatively thick adhesive layer was selected to simulate the situation in civil engineering applications where tolerances need to be compensated in the joints.

The laminates consisted of E-glass fibers embedded in an isophthalic polyester resin. The fiber architecture of both laminate types was similar as they were both composed of complex mat layers on the outside and rovings in the core region. The 12 mm thick laminate comprised two mat layers on each side, while the 6 mm thick laminate contained only one mat layer. A mat layer consisted of a chopped strand mat (CSM) and a woven mat 0°/90° stitched together. A polyester surface veil of 40 g/m² was also applied onto the outer surfaces. To determine the fiber volume fractions, burn-off tests were performed according to ASTM D3171-99.

Table 5.1 Estimated
average layer thickness of 6
and 12 mm laminates

Layer	6 mm GFRP (mm)	12 mm GFRP (mm)
Veil	0.5	0.5
Mat	1.0	0.5 + 1.0
Roving	3.0	8.0

The veil layer totally decomposed, while the fibers of the mat and the roving layers could easily be separated and weighed. The thicknesses of all layers were estimated using an optical microscope and the resulting average values are given in Table 5.1 [13]. The determined total glass fiber volume fractions were 43.6 and 48.5% for 6 mm and 12 mm thick laminates, respectively. Due to the higher fiber content, the mean tensile strength and Young's modulus of the thicker laminates (355 MPa and 34.4 GPa) were higher than those of the 6 mm laminates (283 MPa and 31.4 GPa) [13].

A two-component epoxy system was used (SikaDur 330 from Sika AG) that exhibited an almost elastic behavior and a brittle failure under axial quasi-static tensile loads at ambient temperature. The measured mean tensile strength was 38.1 MPa, the Young's modulus 4.6 GPa and strain to failure 1.0% [14].

All specimens were manufactured and cured in ambient laboratory conditions (23 ± 5°C, 50 ± 10% RH, for ten days). Tests were carried out on an INSTRON 8800 universal testing rig of 100 kN capacity under load control. An environmental chamber was used to control temperature and humidity during testing. Deviations of approximately ± 1°C were recorded for the temperature, while ± 2% differences in relative humidity were observed. Frequency was kept constant at 10 Hz for all joints, while the stress ratio ($R = \sigma_{min}/\sigma_{max}$) was equal to 0.1, resulting in a tension-tension fatigue loading. The frequency of 10 Hz was chosen as a compromise between testing time and hysteretic heating effects. Four different load levels were predetermined for each condition (after an iterative pre-study) to collect experimental data in the range between 10^2 and 10^7 cycles.

The tests were performed under four different controlled environmental conditions: a temperature of -35 ± 1°C (humidity cannot be controlled for negative temperatures), a temperature of 23 ± 1°C and relative humidity of 50 ± 2%, a temperature of 40 ± 1°C and relative humidity of 50 ± 2% and finally, a temperature of 40 ± 1°C and relative humidity of 90 ± 2%. The temperature and humidity ranges were selected in accordance with the properties of the adhesive (which has a glass transition temperature of approximately 50°C) and the operational conditions of the joints as parts of engineering structures. Prior to testing, the specimens were placed inside the chamber for an appropriate time period (approximately 90 min for temperatures above zero and 150 min for the negative temperature) in order to attain the predetermined temperature and humidity levels. Special preconditioning was required for the specimens that were tested at high temperature and high relative humidity. Preliminary quasi-static tests showed that moisture absorption was initially rapid and reached saturation after 70 days. The ultimate load of the joints decreased with increased moisture concentration and

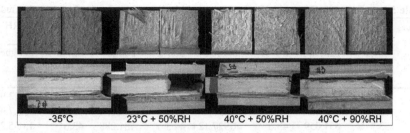

Fig. 5.2 Failure modes under different environmental conditions

also reached a plateau after 70 days. Based on these tests, the fatigue specimens were preconditioned for 70 days in a warm water bath at a temperature of 40°C.

Four different load levels were predetermined for each condition (after an iterative preliminary study) in order to collect experimental data in the range between 10^2 and 10^7 cycles. This range can be considered as representative of a real structure, which may be subjected to 10 million cycles during its lifetime.

5.3 Experimental Results and Analysis

5.3.1 Failure Modes and Fatigue Data

Figure 5.2 shows typical detailed views of failed specimens under different environmental conditions. As it can be seen, all joints exhibited a typical fiber-tear failure: a crack propagated in the region of the interface, with the majority of the cracks located inside the inner GFRP laminate. As reported in [15], fiber-tear failure in pultruded GFRP profiles at ambient temperature occurs because the adherend through-thickness material strength is lower than the adhesive/adherend interfacial strength. This conclusion corresponds with the joint design recommendation according to which joint proportions must be such that the bond never constitutes the weak link in the structure. The strength of the adhesive bond across the bonded area must be greater than the strength of the adherends outside the joint [16].

The crack opening was visible during fatigue loading under all environmental conditions. The dominant failure mechanism and the failure process were similar to those exhibited under quasi-static loading, as presented in [17], although for quasi-static loading it was not possible to observe the damage development, since failure occurred very suddenly without warning or even the prior appearance of any visible cracks. Temperature (in the range between −35 and 40°C) had no noticeable influence on the failure process of the joints; cracks propagated through the mat layers of the 12 mm laminate in all the studied cases. The presence of 90% RH, however, shifted the crack from the adherend to the interface (adhesive crack).

Table 5.2 Fatigue data for double-lap joints (actual measured load level values are reported)

Spec. No.	−35°C		23°C + 50% RH		40°C + 50% RH		40°C + 90% RH	
	Stress level (% of f_u)	Fatigue life	Stress level (% of f_u)	Fatigue life	Stress level (% of f_u)	Fatigue life	Stress level (% of f_u)	Fatigue life
1	53.7	191,251	45.0	2,215,704	43.7	1,909,376	43.7	1,313,940
2	53.4	1,266,390	43.9	1,030,600	47.0	1,667,768	43.6	553,739
3	53.7	651,059	43.6	1,927,124	44.3	484,288	43.6	1,055,244
4	63.6	259,708	53.5	432,844	53.4	149,005	53.4	145,871
5	63.3	85,879	52.7	48,180	53.8	124,173	53.7	96,773
6	63.7	55,178	52.9	27,647	54.1	131,288	53.8	73,037
7	73.3	6,355	62.5	19,788	63.2	17,361	63.2	5,751
8	73.3	23,765	63.5	24,306	63.4	6,731	62.7	10,446
9	73.3	35,569	64.4	53,664	63.5	25,723	63.9	7,905
10	83.0	6,513	76.0	823	73.3	8,861	73.5	1,780
11	82.3	7,171	75.9	207	73.6	9,797	72.2	1,124
12	82.9	2,060	78.0	4,731	73.2	5,813	72.1	1,105

Table 5.3 Estimated statistical parameters for examined fatigue data and static strength

Parameter	−35°C	23°C + 50% RH	40°C + 50% RH	40°C + 90% RH
α_f	22.82	15.57	17.40	24.02
β	72.72	76.67	76.12	64.15
G	0.080	0.081	0.088	0.074
C	0.0022	0.0106	0.0158	0.1520
f_u (MPa)	72.2 ± 3.5	77.2 ± 6.2	70.0 ± 8.8	59.7 ± 4.3

The aforementioned environmental influence is reflected in the collected fatigue data listed in Table 5.2.

All fatigue datasets were subjected to statistical analysis according to the wear-out model described in Chap. 3, and the estimated parameters for all fatigue datasets (and ultimate strength f_u [17]) are listed in Table 5.3.

The F-N curves of the joints under each loading condition can be simulated by substituting the estimated parameters in the following equation, for any level of reliability, $P_S(N)$, (see Chap. 3 for details regarding the process):

$$\sigma_{max} = \beta \left[-LnP_S(N)^{\frac{1}{\alpha_f}} \right] [(N - A)C]^{-G} \qquad (5.1)$$

where $A = -(1-C)/C$.

The entire set of fatigue data is presented in Fig. 5.3. S–N curves based on the wear-out model were derived and plotted (as lines) with experimental fatigue data (symbols) for better visualization and comparison between the four different loading cases. The fatigue life of the examined joints was longer at low temperatures, even if the static strength of the joints at low temperature was lower than the corresponding value under ambient conditions [17]. Analysis showed that,

Fig. 5.3 Derived S-N curves
for examined fatigue datasets

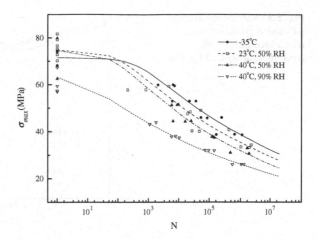

irrespective of loading condition, all curves have similar slopes with an average of
0.081 and a very low standard deviation of 0.005. The increased temperature (from
ambient to 40°C) seems to influence the fatigue life by a factor of approximately
5% for low-cycle fatigue (up to 1,000 cycles), while this difference increases to
10% for longer lifetime. The presence of humidity is even more critical as the
derived S–N curve is situated lower than all the others in the S–N plane, although
it is less steep. Compared to the life of the specimens tested at 40°C and 50% RH,
a significant decrease of roughly one decade of life can be observed when 90% RH
is present.

5.3.2 Modeling Fatigue Life Based on Stiffness Degradation

Stiffness fluctuations during fatigue life were also recorded since stiffness can be
used as a non-destructive damage metric for evaluating the structural integrity of
constructions. The development of phenomenological models, capable of
describing gradual stiffness degradation in terms of macroscopically measured
properties, permits the establishment of fatigue design allowables, which can
easily be incorporated into design codes [18].

During fatigue life, the peak and valley forces and corresponding displace-
ments, Δl, were recorded for each cycle. The secant structural modulus was calculated as:

$$E(N) = \frac{F_{max}(N) - F_{min}(N)}{\Delta l_{max}(N) - \Delta l_{min}(N)} \quad (5.2)$$

where the subscripts max and min indicate the maximum and minimum values of
F and Δl of the N th cycle. Normalized values for number of cycles and stiffness
were used for comparison with the available stiffness data. The initial fatigue

Fig. 5.4 Normalized average
stiffness degradation vs.
normalized fatigue life
(specimens at 40°C, 50% RH)

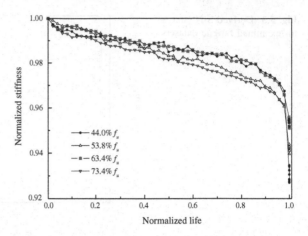

Table 5.4 Residual stiffness at fatigue failure of DLJs under different environmental conditions

−35°C		23°C + RH 50%		40°C + RH 50%		40°C + RH 90%	
Av. stress level (% of f_u)	$E(N)/E(1)$ at failure (%)	Av. stress level (% of f_u)	$E(N)/E(1)$ at failure (%)	Av. stress level (% of f_u)	$E(N)/E(1)$ at failure (%)	Av. stress level (% of f_u)	$E(N)/E(1)$ at failure (%)
53.6	95.5	44.2	94.3	44.0	92.5	43.6	92.8
63.5	96.0	53.0	95.0	53.8	93.8	53.7	94.0
73.3	96.1	63.5	95.9	63.4	95.0	63.3	92.7
82.7	96.8	76.6	95.9	73.4	94.7	72.6	94.3
Average	96.1 ± 0.5	–	95.3 ± 0.8	–	94.0 ± 1.1	–	93.4 ± 0.8

modulus $E(1)$ was calculated using Eq. 5.2 for the first cycle of each test. All stiffness measurements were then normalized with respect to $E(1)$. The number of cycles was divided by the number of cycles to failure.

Basically stiffness degradation results from crack propagation and degradation of laminate stiffness. However, previous experiments on similar adherends [19] showed that degradation of laminate stiffness is insignificant at the low stress levels applied here and, therefore, that degradation can primarily be attributed to crack propagation.

Stiffness degradation was low, with less than 6–7% being observed for all the investigated cases. As shown in Fig. 5.4, for 40°C with 50% RH case, the longer part of the curves exhibits a linear trend, although there are short regions at the beginning and end of life where stiffness changes rapidly. This phenomenon can be attributed to the failure process itself. The damage developed gradually (which can be explained by the stable crack propagation [20]) until close to failure where rapid changes in stiffness indicated unstable crack development. The same trend was exhibited by the specimens tested under other environmental conditions. Nevertheless, increased temperatures were found to cause more fatigue damage, which is reflected, albeit slightly, in more stiffness degradation while the addition of humidity aggravated damage development. As presented in Table 5.4, averaged

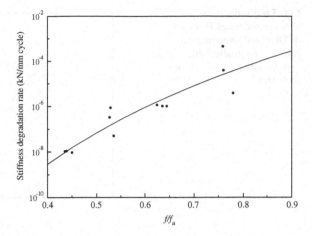

Fig. 5.5 Stiffness degradation rate (absolute value) of DLJs at different load levels

stiffness values at failure are lower for higher testing temperatures and increased humidity levels.

The derivation of stiffness-based fatigue design allowables will be presented in the following paragraphs for the specimens tested under ambient environmental conditions for demonstration. The same process can also be followed for the other datasets.

As shown in Fig. 5.4, stiffness degradation of the examined joints can be considered as linear up to failure, and therefore the following model was established:

$$\frac{E(N)}{E(1)} = 1 - k_3 \left(\frac{f}{f_u}\right)^{k_4} N \tag{5.3}$$

where f denotes the applied stress level and could correspond to its amplitude, maximum value or a normalized value of it. Throughout this work, f_u denotes the maximum applied cyclic load and f_u corresponds to the ultimate tensile load of the joint. Model parameters k_3 and k_4 are dependent on available experimental data for stiffness degradation and it is assumed that they depend on the number of stress cycles and level of the applied load.

Equation 5.3 also establishes a stiffness-based design criterion since for a preset value of stiffness degradation, $E(N)/E(1) = p$, N can be solved to obtain an alternative form of the S–N curve, corresponding not to material failure but to a specific stiffness degradation percentage:

$$N = \frac{E(N) - E(1)}{E(1)k_3 \left(\frac{f}{f_u}\right)^{k_4}} \tag{5.4}$$

For a given specimen, the residual stiffness is assumed to follow Eq. 5.3 with the term $k_3(f/f_u)^{k_4}$ representing the rate of stiffness degradation and assumed to depend on the applied load level. Model parameters are estimated by plotting the stiffness

Fig. 5.6 Comparison between predicted F-N curve of DLJs and experimental results, design allowable corresponding to 2% stiffness reduction

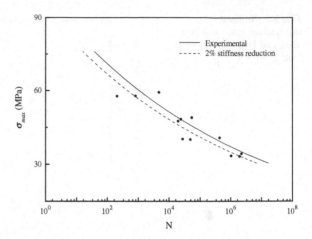

degradation rate against the relevant load levels for all the available experimental data, i.e., 12 specimens for the DLJs, as presented in Fig. 5.5. The resulting estimations of parameters k_3 and k_4 are: $k_3 = -0.00126$ and $k_4 = 14.176$ [21].

After derivation of the model parameters k_3 and k_4, the expected S–N behavior can be simulated by means of Eq. 5.4. The results are presented in Fig. 5.6 and compared to the experimentally determined S–N data. As shown in Fig. 5.6, the theoretical predictions based on the linear model compare well with experimental data. The slight overestimation of fatigue life is attributed to the ignorance of the initial and final periods of stiffness degradation. However, for the DLJs, the effect of these two periods on the entire life is almost negligible. In addition to the F-N curves, Sc-N curves corresponding to predetermined stiffness reduction and not to failure data can be plotted and used as design allowables. For DLJs where total stiffness degradation at failure was less than 7%, the Sc-N curve for 2% decrease of stiffness is plotted, based on the linear model.

5.3.3 Modeling of Fatigue Life with Genetic Programming

The genetic programming modeling process, previously presented in Chap. 4, is also used in this chapter for modeling the fatigue life of the examined adhesively-bonded FRP joints under the different environmental conditions.

In the context of this chapter, the fatigue data of the pultruded joints were treated as follows: all fatigue data except those recorded at 40°C/50% RH were used for the training of the model. A total of 36 fatigue data points were therefore available for the training set. A training dataset was created by using the maximum applied cyclic stress, σ_{max}, the testing temperature, T and the relative humidity RH as input variables. The number of cycles to failure corresponding to each set of input parameters, N_f, was defined as the only output. The training file contained the

Fig. 5.7 Modeling accuracy
of GP

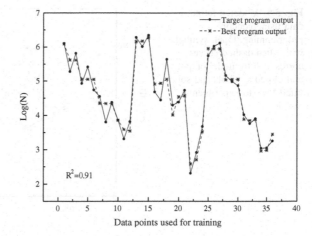

Fig. 5.8 Predicting accuracy
of GP

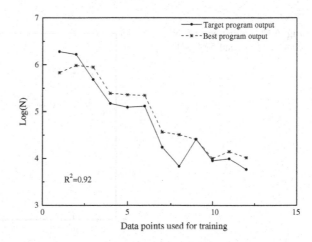

data that the tool used for learning, i.e., the fitness function was calculated using the training file. Given the number of input and output parameters in the training set, the process is characterized as a non-linear stochastic regression analysis. During the training phase, the genetic programming tool established several relationships (by regression analysis) in the form of computer programs between the input and the output variables. Using an iterative process the parameters of the established relationships were adjusted in order to minimize the difference between the theoretical and the real outputs. The same set of data was used for validation of the modeling.

An applied, or test, dataset was subsequently constructed in order to evaluate the predictive ability of the best evolved program after the training process. The applied dataset contains input data for which the output will be calculated by the best evolved program, here the data for the 40°C/50% RH loading case. The same

Fig. 5.9 Modeled fatigue
data using genetic
programming. Open symbols
with fitted dashed curves
correspond to model output
and closed symbols and solid
fitted lines to experimental
results

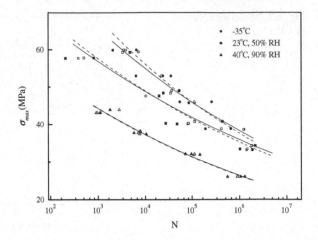

Fig. 5.10 Prediction of S-N
curve at 40°C and 50% RH

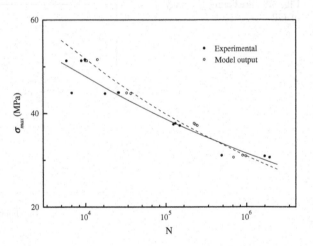

model (the selected evolved program) can be stored and potentially used to predict
other output values for a new applied input dataset.

The training efficiency of the genetic programming tool was very good. As
shown in Fig. 5.7 where target output is compared with the best program output
after the training process, the coefficient of multiple determination (R^2) was
0.91.

The same comment also applies to the predictive efficiency of the emerged GP
model. As presented in Fig. 5.8, the predictive ability of the developed model is
excellent, presenting an R^2 value of 0.92.

This excellent modeling and predictive accuracy of the developed model is also
reflected in the derived S–N curves (Fig. 5.9) for the modeling and fatigue life
prediction of the dataset "unseen" during training (Fig. 5.10).

This study proved the ability of novel computational tools to model and predict the fatigue life behavior of adhesively-bonded joints under different thermomechanical conditions.

5.4 Conclusions

The fatigue behavior of adhesively-bonded pultruded GFRP double-lap joints subjected to a constant amplitude fatigue loading under different environmental conditions has been examined. The influence of temperature and humidity on the damage accumulation and fatigue life of the examined specimens was evaluated. The following conclusions can be drawn:

- Environment had a considerable effect on the fatigue behavior of the examined joints. Increased temperature seemed to shorten specimen fatigue life. This phenomenon is more pronounced in the presence of high humidity levels. The derived S–N curves had almost the same slope, which indicates that joint fatigue behavior is very much dependent on the joint's static strength.
- When tested at −35, 23 and 40°C, cracks leading to the failure of the joints propagated through the outer layer of the pultruded adherends. However, in the presence of high humidity, the failure shifted from an adhesive to an interfacial failure, which implied that the adhesive/composite bond was weaker than the fiber/matrix bond in the composite itself.
- Almost linear stiffness degradation was observed up to failure irrespective of the applied load level. Very close to failure, a rapid decrease in stiffness was observed for all the investigated cases. An increase in temperature was found to provoke higher stiffness degradation, aggravated by the addition of humidity. However, only a small stiffness degradation of 6–7% was measured for all the specimens prior to ultimate failure.
- The fatigue life of the examined joints can be modeled either by the traditional S–N curves, based on stiffness degradation measurements, or using novel computational tools such as genetic programming.
- Although modeling based on stiffness degradation measurements requires more effort than the simpler stress-based approach (in terms of equipment, complicated recording set-up and/or calculations) it has the merit of also being able to specify allowable stiffness reduction levels. Moreover, since these methods are based on stiffness measurements that can be performed during the operational life of structures without interruptions and in a non-destructive manner, they can be adapted by design codes as on-line health monitoring tools.
- Modeling of the fatigue behavior of the examined structural components with genetic programming was proved feasible. It was also shown that this kind of tool can be used for prediction of the behavior of composite materials and structural components, like the adhesively-bonded joints examined in this chapter, under complex thermomechanical loading conditions, where other methods are unable to derive reliable results.

References

1. R.B. Gilmore, S.J. Shaw, The effect of temperature and humidity on the fatigue behavior of composite bonded joints. Composite Bonding, ASTM STP 1227, (1974)
2. I.A. Ashcroft, D.J. Hughes, S.J. Shaw, Adhesive bonding of fibre reinforced polymer composite materials. Assembly Autom. **20**(2), 150–161 (2000)
3. I.A. Ashcroft, D.J. Hughes, S.J. Shaw, M.A. Wahab, A. Crocombe, Effect of temperature on the quasi-static strength and fatigue resistance of bonded composite double lap joints. J. Adhes. **75**, 61–68 (2001)
4. J.A.M. Ferreira, P.N. Reis, J.D.M. Costa, M.O.W. Richardson, Fatigue behaviour of composite adhesive lap joints. Compos. Sci. Technol. **62**(10–11), 1373–1379 (2002)
5. Y. Miyano, M. Nakada, R. Muki, Prediction of fatigue life of a conical shaped joint system for reinforced plastics under arbitrary frequency, load ratio and temperature. Mech. Time Depend Mat. **1**, 143–159 (1997)
6. I. Malvade, A. Deb, P. Biswas, A. Kumar, Numerical prediction of load-displacement behaviors of adhesively bonded joints at different extension rates and temperatures. Comp. Mater. Sci. **44**(4), 1208–1217 (2009)
7. Y. Miyano, M. Nakada, H. Kudoh, R. Muki, Prediction of tensile fatigue life for unidirectional CFRP. J. Compos. Mater. **34**(7), 538–550 (2000)
8. A.P. Vassilopoulos, E.F. Georgopoulos, V. Dionyssopoulos, Modelling fatigue life of multidirectional GFRP laminates under constant amplitude loading with artificial neural networks. Adv. Compos. Lett. **15**(2), 43–51 (2006)
9. A.P. Vassilopoulos, E.F. Georgopoulos, V. Dionyssopoulos, Artificial neural networks in spectrum fatigue life prediction of composite materials. Int. J. Fatigue **29**(1), 20–29 (2007)
10. A.P. Vassilopoulos, E.F. Georgopoulos, T. Keller, Comparison of genetic programming with conventional methods for fatigue life modelling of FRP composite materials. Int. J. Fatigue **30**(9), 1634–1645 (2008)
11. A.P. Vassilopoulos, R. Bedi, Adaptive neuro-fuzzy inference system in modelling fatigue life of multidirectional composite laminates. Comp. Mater. Sci. **43**(4), 1086–1093 (2008)
12. A.P. Vassilopoulos, T. Keller, Modeling of the fatigue life of adhesively-bonded FRP joints with genetic programming. 17th International Conference on Composite Materials (ICCM17), Edinburgh, UK, 27–31 July (2009)
13. Y. Zhang, T. Keller, Progressive failure process of adhesively bonded joints composed of pultruded GFRP. Compos. Sci. Technol. **68**(2), 461–470 (2008)
14. J. De Castro, T. Keller, Ductile double-lap joints from brittle GFRP laminates and ductile adhesives. Part I: Experimental investigation. Compos. B Eng. **29**(2), 271–281 (2008)
15. T. Keller, T. Vallée, Adhesively bonded lap joints from pultruded GFRP profiles, Part I: Stress-strain analysis and failure modes. Compos. B Eng. **36**(4), 331–340 (2005)
16. L.J. Hart-Smith, The key to designing durable adhesively bonded joints. Composites **25**(9), 895–898 (1994)
17. Y. Zhang, A.P. Vassilopoulos, T. Keller, Effects of low and high temperatures on tensile behavior of adhesively-bonded GFRP joints. Compos. Struct. **92**(7), 1631–1639 (2010)
18. J.R. Gregory, S.M. Spearing, Constituent and composite quasi-static and fatigue fracture experiments. Compos. A Appl. S. **36**(5), 665–674 (2005)
19. T. Keller, T. Tirelli, A. Zhou, Tensile fatigue performance of pultruded glass fiber reinforced polymer profiles. Compos. Sci. Technol. **68**(2), 235–245 (2005)
20. Y. Zhang, A.P. Vassilopoulos, T. Keller, Environmental effects on fatigue behavior of adhesively–bonded pultruded structural joints. Compos. Sci. Technol. **69**(7–8), 1022–1028 (2009)
21. Y. Zhang, A.P. Vassilopoulos, T. Keller, Stiffness degradation and fatigue life prediction of adhesively–bonded joints for fiber–reinforced polymer composites. Int. J. Fatigue **30**(10–11), 1813–1820 (2008)

Chapter 6
Macroscopic Fatigue Failure Theories for Multiaxial Stress States

6.1 Introduction

A considerable number of fatigue theories and methodologies for the fatigue life prediction of composite materials and structures have been developed, based on empirical, phenomenological modeling or on the quantification of specific damage metrics, such as the residual strength and/or stiffness of the examined material or structural element. Structural elements made of composite materials were used to be treated as being subjected to uniaxial stress states comprised by the maximum developed stress tensor component, whereas the other components were neglected. This assumption seems reasonable for the highly anisotropic composite materials and was even adopted by the scientific community in the past. For example, a composite wind turbine rotor blade was treated by state-of-the-art design codes, e.g., [1, 2], as a typical beam-like structure for which fatigue life calculations are limited in that they consider only the action of the normal stress component in the beam axis direction. However, regardless of the effect of various model parameters on the accuracy of the theoretical predictions, a strong effect of the transverse normal and shear stress components that develop during service life on the fatigue life has been recorded, e.g., [3]. Theories that do not take into account the interaction of the different stress tensor parameters, like the maximum stress theory (Rankine) or the maximum normal strain theory (Saint Venant) are reliable only for specific cases of isotropic materials e.g. [4], where it is claimed that the maximum stress theory is the most appropriate for isotropic materials that fail due to a brittle fracture, while the maximum shear stress theory or the maximum distortional energy theory seems the most appropriate for the prediction of ductile material behavior. The maximum stress criterion for orthotropic laminae was apparently first suggested in 1920 by Jenkins [5] as an extension of Rankine's theory for isotropic materials. The criterion predicts failure when any principal material axis stress component exceeds the corresponding strength. As shown in [5], since the strengths along the principal material directions provide the input to the criterion, the agreement between the

A. P. Vassilopoulos and T. Keller, *Fatigue of Fiber-reinforced Composites*,   155
Engineering Materials and Processes, DOI: 10.1007/978-1-84996-181-3_6,
© Springer-Verlag London Limited 2011

theoretical failure surfaces and the experimental biaxial failure data for a unidirectional graphite/epoxy composite is fair when the applied stress is uniaxial along these directions. However, due to the lack of stress interaction, the accuracy of the predictions is lower in biaxial stress situations.

Nevertheless, quadratic interaction failure criteria that take into account the effect of the different stress tensor parameters on the strength or the fatigue life of the examined materials also exist. One of the first theories of this type was introduced by von Mises, also known as the maximum distortional energy criterion, around 1900 [5]. It is the most widely used quadratic interaction criterion for predicting the onset of yielding in isotropic materials. A generalization of von Mises' theory to incorporate the anisotropic behavior of initially isotropic metals exhibited by the material during large plastic deformations was established by Hill in 1948. This criterion has the drawback that it does not consider the different strength in tension and compression, or even the different positive or negative shear strength that is exhibited by some types of composite materials. One way to take into account the different strength is to include terms which are linear with respect to the normal stresses (to the power of 1) as suggested by Hoffman [6]. This phenomenon, also called *differential failure*, is accommodated by the Tsai-Wu failure criterion introduced in 1971, [7].

It is obvious that a significant number of multiaxial failure theories exist and therefore the selection of the most appropriate one for the examined material and loading conditions is left to the judgment of the design engineer.

The fatigue life prediction is much more complicated since material properties deteriorate during loading. An additional difficulty is caused by the fact that this deterioration is not linear, but depends on the loading conditions and history of loading, i.e., the degree of damage already caused to the material due to previous loading. The inability to simulate material behavior leads to the adoption of high safety factors, which, together with the safety factors due to the stochastic nature of the fatigue phenomenon, result in the overdesign of any structure. New damage-tolerance design philosophies make fatigue theories able to reliably predict the fatigue behavior of a composite material under different loading conditions compulsorily.

Predictive formulations of the fatigue life of composite materials which take stress multiaxiality into account appeared in the 1970s. Most of the proposed criteria were generalizations of static criteria in order to take into account fatigue parameters, such as number of cycles to failure, frequency and stress ratio. Both uniaxial and multiaxial fatigue experiments were performed to assist the development of the fatigue theories and evaluate their predictive ability. Hashin and Rotem [8] proposed a fatigue strength criterion for fiber-reinforced materials based on the different failure modes exhibited. According to the authors, two failure modes exist for unidirectional materials, the fiber and the matrix failure modes. When multidirectional laminates are considered, another failure mode, the interlaminar, is encountered [9]. The Hashin—Rotem failure criterion can be implemented only for materials for which failure modes can be clearly discriminated.

Owen and Griffiths [10] presented a static and fatigue experimental program of glass/polyester thin-walled tubes under combined axial loading and internal pressure in order to evaluate existing multiaxial failure criteria. The authors compared the predictions of the examined failure criteria with the obtained fatigue data and concluded that only those theories that involve complex stress properties provide a reasonable fit. Fujii and Lin [11], in order to validate their experimental, biaxial tension/torsion fatigue data, adapted the Tsai-Wu strength criterion [7] for fatigue. However, the adaptation proved to be insufficient to correctly predict the fatigue behavior of the investigated material system for all the considered fatigue lives. The authors replaced the tensile static strength parameters used in the original version of the Tsai-Wu failure criterion by corresponding S–N curves. However, since they did not have experimental fatigue data under compressive loading they decided to fit the modified Tsai-Wu equation to the available fatigue data for a selected number of cycles in order to estimate the compressive fatigue strength in terms of S–N curves. Therefore, they proved that the proposed form of the Tsai-Wu criterion can accurately model the fatigue behavior of the examined material for certain numbers of cycles–those selected for the fitting of the unknown fatigue strengths – but it fails to predict the behavior for longer fatigue lifetimes.

Sims and Brogdon [12] developed a fatigue failure criterion based on the Tsai-Hill theory. They replaced the static strengths in the Tsai-Hill theory by corresponding fatigue functions. A comprehensive fatigue program comprising a number of tension-tension and interlaminar shear fatigue experiments on S-glass/epoxy and graphite/epoxy unidirectional and cross-ply laminates was performed in order to assist the development of the fatigue theory. Fatigue data on longitudinal, transverse and $\pm 45°$ laminates provided the necessary information to determine the in-plane shear fatigue function, which has been proved significantly different from the interlaminar shear fatigue function. The fatigue strength of any other off-axis laminates can then be obtained through the use of the principal fatigue functions and the established theoretical model. The authors concluded that the proposed theory can be useful for the preliminary fatigue design of fatigue-critical components since it requires limited fatigue data. However, the approach was proved conservative since only the first-ply failure of the examined laminates could be accurately predicted.

Jen and Lee [13, 14] attempted to predict the fatigue behavior of off-axis loaded AS4 carbon/PEEK APC2 unidirectional thermoplastic composite laminates by introducing an appropriately modified version of the Tsai-Hill criterion. They extended the predictive capability of their theoretical formulation for multidirectional laminates made of the same prepreg by means of classical lamination theory (CLT) and ply-discount considerations. Based on the CLT and the assumed ply-discount method for the stiffness degradation of a failed ply, their theoretical predictions for the examined thermoplastic laminates were satisfactorily corroborated by experimental evidence.

An extension of the quadratic version of the Failure Tensor Polynomial, as interpreted by Tsai and Hahn [15], for the prediction of fatigue strength under complex stress states was introduced in [16, 17] by Philippidis and Vassilopoulos.

This strength criterion, established on the basis of the fatigue data presented in Chap. 2, was shown to yield reliable predictions when compared to experimental data from constant amplitude (CA) biaxial fatigue experiments on a wide variety of composite laminates. Satisfactory predictions were also produced using the criterion for the static strength of the GFRP laminate under investigation by testing on- and off-axis specimens in tension and compression. The approach differs from previously mentioned studies, since a direct characterization approach was adopted in [16, 17] for manipulation of the fatigue data and the life prediction of the examined $[0/(\pm 45)_2/0]_T$ laminate. The material was considered a "homogeneous" anisotropic continuum, thus avoiding uncertainties in the modeling of stiffness degradation of failed layers or interlaminar effects and interactions. This is particularly useful for the investigated GFRP laminate, which comprised different glass fabrics with stitched fibers at various orientations. In such cases, simplistic theoretical considerations for load distribution and ply stiffness degradation are no longer applicable.

Kawai [18] examined the off-axis fatigue behavior of unidirectional CFRP composites and developed a fatigue damage mechanics model that could take into account the off-axis angle and stress ratio effect under any constant amplitude loading with non-negative mean stresses. A non-dimensional effective stress parameter was determined by dividing the maximum applied stress at any off-axis angle by the corresponding static strength. The Tsai-Hill criterion was used to calculate static strength at different off-axis angles. The results showed that the model is capable of adequately predicting the off-axis fatigue behavior of unidirectional GFRP and CFPR laminates over a range of non-negative mean stresses.

Fawaz and Ellyin [19] proposed a multiaxial fatigue failure criterion based on only one experimental S–N curve, and the material's static strengths. Multiaxiality is introduced via static failure criteria and S–N curves under complex stress states can be predicted. This criterion constitutes a flexible fatigue failure condition, capable of yielding satisfactory predictions. Wide acceptance is, however, restricted by its high sensitivity to the choice of the single experimental S–N formulation required for its application, as has already been proved elsewhere [17, 20].

The fatigue life prediction of composite laminates has also been addressed from a different perspective, based on the strain energy concept, for the development of a general fatigue failure criterion to take the stress multiaxiality, stress ratio and frequency into account. El Kadi and Ellyin [21] introduced a fatigue failure criterion for unidirectional composite laminates based on the input strain energy that considers both fiber orientation and stress ratio. They found that a normalized form of the criterion can mutate all the fatigue data to fall onto a single line, however, presenting significant scatter. Shokrieh and Taheri [22] developed a fatigue failure model for unidirectional polymer composite laminates based on the static strain energy failure criterion presented in [23]. They also found that use of their theory derives a "master" curve that can adequately model the fatigue behavior of the examined unidirectional material system under any constant amplitude loading pattern.

A review of the multiaxial fatigue theories for composite laminates was recently presented by Quaresimin et al. [20]. As mentioned in [20], although some of the existing theories are accurate, they cannot always guarantee a safe fatigue design under any loading condition. However, the fairly simple formulation of some of the criteria and their straightforward application make them attractive for design procedures and implementation in numerical codes.

The aforementioned literature review revealed that several methods have been proposed to address the problem of fatigue life prediction of composite materials under multiaxial stress states. A comparison of the predictive ability of the selected methods is presented in this chapter. Fatigue failure theories representative of most of the existing concepts are briefly reviewed and their predictive ability is assessed. The fatigue life prediction performance of the examined fatigue failure criteria will be evaluated on the basis of their capacity to predict the off-axis fatigue behavior of the material system examined in Chap. 2 of this book and other composite material systems from the literature, loaded under uniaxial and biaxial fatigue loads.

6.2 Fatigue Failure Theories

Six multiaxial fatigue failure theories are reviewed in this chapter and their applicability and predicting ability are evaluated. Three of the selected models can be classified as macroscopic fatigue strength criteria, which are usually generalizations of known static failure theories to take into account factors relevant to the fatigue life of the structure such as number of cycles and loading frequency: the Hashin-Rotem (HR) [8], Sims-Brogdon (SB) [12] and the Failure Tensor Polynomial in Fatigue (FTPF) [16, 17]. The applicability of these theories is based on a dataset containing at least three experimentally derived S–N curves, and the static strengths of the material for the HR criterion. Two more macroscopic failure criteria proposed by Kawai (KW) [18] and Fawaz and Ellyin (FE) [19] include the stress ratio in their formulation and can thus be implemented for the prediction of the fatigue life of a composite material for a wide range of loading patterns based on limited databases. The sixth model, introduced by Shokrieh and Taheri (ST) [22], is based on a strain energy concept. The last three criteria are based on one fatigue dataset used for estimation of the model parameters and static strengths of the material that are used in combination with any valid static failure criteria for calculation of off-axis strengths. The S–N curve used is designated "reference" or "master" curve and the methods can be designated "master curve methods". Other models can be found in the literature as well, but those selected here were proved more accurate, e.g. see [16, 20]. They can be easily implemented, and sufficient fatigue data exist in the literature for evaluation of their accuracy.

6.2.1 Hashin-Rotem

One of the first attempts to generalize a static failure theory in order to take into account factors relevant to fatigue was made by Hashin and Rotem (HR) [8].

The authors presented a fatigue failure criterion based on the different damage modes demonstrated during failure. According to the authors, there are two main modes in the case of unidirectional materials: the fiber failure mode and the matrix failure mode. The discrimination between these two modes is based on the off-axis angle of the reinforcement in relation to the loading direction. The critical fiber angle, which defines the transition from one failure mode to another, is related to the strengths of the material and can be estimated by the following equation:

$$\tan \theta_c = \frac{\tau^s \, f_\tau(R,N,fr)}{\sigma_A^s \, f_A(R,N,fr)} \tag{6.1}$$

where τ^s and σ_A^s are the static shear and longitudinal (axial) strengths, respectively, while the functions $f_\tau(R,N,fr)$, and $f_A(R,N,fr)$ are the fatigue functions of the material along the same directions, and are related to the stress ratio, $R = \sigma_{min}/\sigma_{max}$, number of cycles, N and fatigue frequency, fr. The S–N curves of the material under shear, τ, and under longitudinal, σ_A, or transverse, σ_T, directions are given as the product of the static strengths along any direction and the corresponding fatigue function.

If the reinforcement forms an angle of less than θ in relation to the loading direction, the fiber mode is the prevailing failure mode, otherwise the matrix failure mode leads to fatigue failure. Therefore the failure criterion has two forms:

$$\sigma_A = \sigma_A^u \tag{6.2a}$$

$$\left(\frac{\sigma_T}{\sigma_T^u}\right)^2 + \left(\frac{\tau}{\tau^u}\right)^2 = 1 \tag{6.2b}$$

where superscript u denotes fatigue failure stress or the S–N curve of the material in the corresponding direction and subscript T denotes transverse to the fiber direction. It can be shown that any off-axis fatigue function $f''(R,N,fr)$ (matrix failure mode), can be given as a function of, f_τ, f_T, τ^s, σ_T^s and the angle θ [8]:

$$f''(R,N,fr) = f_\tau \sqrt{\frac{1 + \left(\frac{\tau^s}{\sigma_T^s}\right)^2 \tan^2 \theta}{1 + \left(\frac{\tau^s}{\sigma_T^s}\frac{f_\tau}{f_T}\right)^2 \tan^2 \theta}} \tag{6.3}$$

Equation (6.3) can be used not only for the calculation of any off-axis fatigue function, but also to calculate fatigue functions f_τ and f_T, from two different experimentally obtained off-axis fatigue functions. For the application of this criterion over the entire range of on- and off-axis directions, three S–N curves must be defined experimentally, along with the corresponding static strengths of the material.

When the laminate is multidirectional [9], the case is far more complicated. As each lamina is under a different stress field, failure may occur in some laminae after a certain amount of load cycling while the other laminae are still far from failure. The different stress fields developed in each lamina produce interlaminar stresses, able to cause successive failure. To take these stresses into account, an interlaminar failure mode is established and the set of equations (Eq. 6.2a) is supplemented by:

$$\left(\frac{\sigma_d}{\sigma_d^u}\right)^2 + \left(\frac{\tau_d}{\tau_d^u}\right)^2 = 1 \tag{6.4}$$

where subscript d denotes the interlaminar stress components.

The Hashin and Rotem failure criterion can predict the fatigue behavior of a unidirectional (UD) or multidirectional (MD) laminate subjected to uniaxial or multiaxial cyclic loads provided that the type of failure mode exhibited during fatigue failure can be distinguished. Its use for woven or stitched fabrics is therefore not recommended.

6.2.2 Fawaz-Ellyin

Fawaz and Ellyin (FWE) [19] proposed a fatigue failure criterion to simulate the fatigue behavior of unidirectional and multidirectional composite laminates under multiaxial cyclic stress states. The criterion has the advantage of requiring only one experimentally obtained S–N curve and the static strengths of the laminate along different directions. The idea is based on the assumption that all the on- and off-axis S–N curves of the laminate, when normalized by the corresponding static strengths, lie in a narrow band on the S–N plane. Although this idea had already been proposed in 1981 by Awerbuch and Hahn [24], where a master fatigue curve was demonstrated as being the representative S–N curve of a number of normalized off-axis fatigue datasets, it was Fawaz and Ellyin who developed a failure criterion based on this.

According to this criterion, if a reference S–N curve exists, e.g.,:

$$S_r = m_r log(N) + b_r, \tag{6.5}$$

the S–N curve under any off-axis angle or biaxial ratio can be calculated by:

$$S(a_1, a_2, \theta, R, N) = f(a_1, a_2, \theta)(b_r + g(R)m_r \log(N)) \tag{6.6}$$

as a function of the reference S–N curve.

In Eqs. 6.5 and 6.6, subscript r denotes the reference direction and a_1 and a_2 are the transverse over normal (σ_y/σ_x) and shear over normal (σ_s/σ_x) biaxial stress ratios. Parameters m_r and b_r are derived after fitting to the experimental data, while model functions f and g are non-dimensional and are defined by:

$$f(a_1, a_2, \theta) = \frac{\sigma_x(\alpha_1, \alpha_2, \theta)}{X_r} \tag{6.7}$$

$$g(R) = \frac{\sigma_{max}(1 - R)}{\sigma_{max_r} - \sigma_{min_r}} \tag{6.8}$$

with $\sigma_x(a_1, a_2, \theta)$ being the static strength along the longitudinal direction and X_r the static strength along the reference direction. The off-axis static strengths of the examined material can be estimated using any reliable multiaxial static failure criterion, e.g., Tsai-Hahn. As it can be seen from Eq. 6.8, function g is introduced to take into account different stress ratios, R. When the stress ratio of the reference S–N curve is the same as that of the S–N curve being predicted, $g = 1$, while for $R = 1$ (quasi-static loading), g equals 0.

Although the FWE criterion in Eq. 6.5 was initially presented based on the Lin-Log S–N curve representation (following the original formulation as described in [19]), it is obvious that any other type of S–N curve, e.g., Log-Log, can also be used without loss of the generality of the method.

Although the FWE criterion has the advantage of requiring only a minimum amount of data, the predictions are very sensitive to the selection of the reference curve, as it will be shown in the following.

6.2.3 Sims-Brogdon

Sims-Brogdon (SB) [4] modified the Tsai-Hill failure criterion for static strengths to a fatigue criterion by replacing the static strengths with corresponding fatigue functions. As a result, the Tsai-Hill tensor polynomial in fatigue is:

$$\left(\frac{K_1}{\sigma_1}\right)^2 - \frac{K_1 K_2}{\sigma_1^2} + \left(\frac{K_2}{\sigma_2}\right)^2 + \left(\frac{K_{12}}{\sigma_6}\right)^2 = \frac{1}{\sigma_x^2} \tag{6.9}$$

where σ_i, $i = 1, 2, 6$ denotes the fatigue functions (for the SB criterion the corresponding S–N curves) along the longitudinal, the transverse directions and shear, respectively, and the parameters K_1, K_2 and K_{12} are the ratios of the stresses along the principal material system over the lamina stress in the direction of the load. These parameters are expressions of the cos and sin of the off-axis angle under consideration. The desired laminate fatigue strength at any off-axis angle is designated by σ_x.

This was the first attempt to modify static polynomial failure criteria in order to take fatigue parameters, and especially the number of cycles to failure, into account. However, the criterion refers to lamina fatigue strength and can be extended to laminates of any orientation using laminated plate theory and knowledge of the stresses in the individual lamina to predict first-ply failure. The SB fatigue theory has the same drawback as the Tsai-Hill criterion, which was used as the basis for its development–it does not take the different strengths of the material under tension and compression into account.

6.2.4 Failure Tensor Polynomial in Fatigue

A modification of the quadratic version of the failure tensor polynomial for the prediction of fatigue strength under complex stress states was introduced by Philippidis and Vassilopoulos [16, 17] and designated Failure Tensor Polynomial in Fatigue (FTPF). The theory is based on the Tsai-Hahn tensor polynomial and adapted for fatigue by substituting the failure tensor components with the corresponding S–N curves. Therefore the FTPF criterion can be expressed in the material symmetry axes (1 and 2), under plane stress by:

$$F_{11}\sigma_1^2 + F_{22}\sigma_2^2 + 2F_{12}\sigma_1\sigma_2 + F_1\sigma_1 + F_2\sigma_2 + F_{66}\sigma_6^2 - 1 = 0, \qquad (6.10)$$

with the components of the failure tensors given by:

$$F_{11} = \frac{1}{XX'}, F_{22} = \frac{1}{YY'}, F_{66} = \frac{1}{S^2}, F_1 = \frac{1}{X} - \frac{1}{X'}, F_2 = \frac{1}{Y} - \frac{1}{Y'} \qquad (6.11)$$

where F_{ii} and F_i are functions of the number of cycles, N, stress ratio, R and the frequency, fr, of the loading since the failure stresses have been substituted with the S–N curves. X, Y and S represent the fatigue strengths of the material (being in general functions of the number of cycles, the stress ratio and the fatigue frequency) along the longitudinal, the transverse directions and under shear loading. The prime$'$ is used for compressive fatigue strengths. Simple uniaxial constant amplitude fatigue experiments can be used for the derivation of $X(N, R, fr)$ and $Y(R, N, fr)$. However, the application of pure shear stresses on a composite specimen is more difficult. Since methods like the rail-shear test on plane specimens or torsional tests on cylindrical specimens are quite complicated and costly, alternative methods were developed, e.g., [25, 26].

The method for the derivation of the shear fatigue strength as half of the S–N curve of the specimens cut at 45° off-axis was initially examined in [16], for the material systems presented in Chap. 2, and showed satisfactory results for the case of reversed loading. However, as it was proved later, [17], a value of 1/2.2 for the S–N at 45° off-axis offered more accurate results for the reversed loading and the other loading patterns that were examined. An alternative method was also presented in [17] to avoid this inconvenience. According to this, the shear fatigue strength can be estimated directly by means of the FTPF criterion and by the use of one off-axis fatigue curve. This method for the back-calculation of the fatigue shear strength, which is presented in detail later on, has been proved sufficiently accurate, especially for unidirectional composite laminates, without the problems raised by use of the 45° off-axis divided by a factor of 2, or 2.2.

Although the components of the failure tensor presented in Eq. 6.11 can be easily determined by simple uniaxial experiments, the selection of the interactive term, F_{12}, is much more complicated. The burden of biaxial experiments is significant and in several cases the results are not sufficiently reliable, see for example [27 and 28], for a discussion on this subject. Therefore, the selection of the interactive term is based on experience and geometrical/mathematical considerations:

- The failure locus defined by Eq. 6.10 must include the origin of the co-ordinate system otherwise, the criterion would predict failure even under zero load application.
- It is widely accepted and supported by the experimental evidence that the failure locus must be closed (e.g., ellipse).

The following stability criterion between the failure tensor terms and the interaction term must be met in order to satisfy the aforementioned conditions:

$$F_{11}F_{22} - F_{12}^2 \geq 0 \tag{6.12}$$

or in general:

$$F_{ii}F_{jj} - F_{ij}^2 \geq 0 \tag{6.13}$$

The selection of the form of the interactive term leads to different criteria. The following will be used here:

$$F_{12} = -0.5\sqrt{F_{11}F_{22}} \tag{6.14}$$

although it has been proved, e.g., in [29], that the use of the interaction term in the form:

$$F_{12} = 0.5(F_{33} - F_{11} - F_{22}) \tag{6.15}$$

can provide more accurate predictions. However, the derivation of the out-of-plane strength of the (thin) composite laminates that is required for the application of Eq. 6.15 is a demanding task that outweighs the advantages of the improved predictions, and therefore Eq. 6.14 is preferred.

Substitution of the failure tensor components (Eq. 6.11) and the interaction term (Eq. 6.14) into Eq. 6.10 gives for the plane stress state:

$$\frac{\sigma_1^2}{XX'} + \frac{\sigma_2^2}{YY'} - \frac{\sigma_1\sigma_2}{XY} + \frac{\sigma_6^2}{S^2} - 1 = 0 \tag{6.16}$$

and taking into account the tensorial transformation equations:

$$\sigma_1 = \sigma_x \cos^2\theta$$
$$\sigma_2 = \sigma_x \sin^2\theta \tag{6.17}$$
$$\sigma_6 = \sigma_x \sin\theta\cos\theta$$

the fatigue strength (σ_x) at any off-axis angle can be calculated as:

$$\sigma_x = \sqrt{\frac{1}{\left(\frac{\cos^4\theta}{XX'} + \frac{\sin^4\theta}{YY'} - \frac{\cos^2\theta\sin^2\theta}{XY} + \frac{\cos^2\theta\sin^2\theta}{S^2}\right)}} \tag{6.18}$$

Equation 6.18 can also be used for calculation of the S–N curve under shear, when it is not possible to derive it experimentally. For this, an off-axis S–N curve (σ_x) must be used as the reference curve and Eq. 6.18 must be solved for S.

6.2.5 Kawai

Kawai [18] developed a non-dimensional stress parameter-based theory founded on the Tsai-Hill static failure criterion by introducing a non-dimensional effective stress Σ^* that takes the effects of the stress ratio and the state of stress on the fatigue life into account. The non-dimensional effective stress is given by:

$$\Sigma^* = \frac{\frac{1}{2}(1-R)\sigma^*_{max}}{1 - \frac{1}{2}(1+R)\sigma^*_{max}} \qquad (6.19)$$

Equation 6.19 is valid for $|R| \leq 1$. σ^*_{max} is the maximum non-dimensional effective stress which can be calculated by:

$$\sigma^*_{max} = \Omega(\theta)\sigma_{max}, \qquad (6.20)$$

as a function of the maximum cyclic stress of the developed plane stress state ($\sigma_{max}(>0)$) and the "orientation factor" [18] given by:

$$\Omega(\theta) = \left(\frac{\cos^4\theta}{X^2} - \frac{\sin^2\theta\cos^2\theta}{X^2} + \frac{\sin^4\theta}{Y^2} + \frac{\sin^2\theta\cos^2\theta}{S^2} \right) \qquad (6.21)$$

following the Tsai-Hill failure criterion for quasi-static stress states. The considered fatigue life model is,

$$2N_f = \frac{1}{(\Sigma^*)^{n^*}} \qquad (6.22)$$

where $2N_f$ is the number of reversals to failure and n^* is the material constant obtained by fitting to experimental fatigue data. Equation 6.22 shows that the master S–N relationship obtained by plotting the non-dimensional effective stress Σ^* against the number of reversals to failure on logarithmic scales becomes a straight line.

To apply the model for off-axis fatigue simulation, the S–N relationship for each off-axis angle (θ) can be reproduced from the master S–N curve by the following equation:

$$\sigma_{max} = \frac{2\Sigma^*}{\Omega(\theta)[(1-R) + (1+R)\Sigma^*]} \qquad (6.23)$$

Application of the proposed model to the unidirectional glass/epoxy and carbon/epoxy composites under constant amplitude cyclic loading with non-negative mean stresses [18] showed that the model can adequately describe off-axis fatigue behavior under the examined loading conditions.

6.2.6 Shokrieh-Taheri

Shokrieh and Taheri (ST) [22] proposed a strain energy-based model, derived from the Sandhu static failure energy criterion [23], for predicting the fatigue life of a unidirectional lamina at various fiber angles and stress ratios. They also adopted the assumption of El Kadi and Ellyin [21] that the relationship between fatigue life and total input energy can be described by a power law type equation of the form of Eq. 6.24:

$$\Delta W^t = kN^\alpha + C \tag{6.24}$$

Additionally they assumed $C = 0$ and independency of the fitting parameters, k and α, of the stress ratio and fiber orientation. According to [22] these two parameters can be obtained for each material by using one set of fatigue test data at an arbitrary stress ratio and fiber orientation.

The proposed model, with respect to the on-axis coordinate system, is given by the following equation:

$$\Delta W = \Delta W_I + \Delta W_{II} + \Delta W_{III} = \frac{\Delta\sigma_1\Delta\varepsilon_1}{X\varepsilon_{u1}} + \frac{\Delta\sigma_2\Delta\varepsilon_2}{Y\varepsilon_{u2}} + \frac{\Delta\sigma_6\Delta\varepsilon_6}{S\varepsilon_{u6}} \tag{6.25}$$

where ΔW represents the sum of strain energy densities contributed by all stress components in material directions and Δ before a symbol denotes range, e.g., $\Delta\sigma$ denotes the stress range. W_I, W_{II} and W_{III} denote the strain energy densities in the longitudinal and transverse directions and shear, respectively and can be derived by the set of Eqs. 6.26:

$$\Delta W_I = \frac{1}{X^2}\frac{(1+R)}{(1-R)}(\Delta\sigma_x)^2\cos^4\theta$$

$$\Delta W_{II} = \frac{1}{Y^2}\frac{(1+R)}{(1-R)}(\Delta\sigma_x)^2\sin^4\theta \tag{6.26}$$

$$\Delta W_{III} = \frac{1}{S^2}\frac{(1+R)}{(1-R)}(\Delta\sigma_x)^2\sin^2\theta\cos^2\theta$$

where, in addition to the material static strengths, X, Y and S, σ_1, σ_2, σ_6, and ε_1, ε_2, ε_6 are stress and strain tensor components, while ε_{u1}, ε_{u2} and ε_{u6} are the maximum strains in the principal material directions.

Assuming a linear stress-strain response along the material directions, the conversion of off-axis stresses into the on-axis coordinate Eq. 6.25 takes the form:

$$\Delta W = \frac{(1+R)}{(1-R)}(\Delta\sigma_x)^2 \left(\frac{\cos^4\theta}{X^2} + \frac{\sin^4\theta}{Y^2} + \frac{\sin^2\theta\cos^2\theta}{S^2} \right) \qquad (6.27)$$

with θ being the fiber orientation angle. Equation 6.27 is valid as long as $R \geq 0$.

The advantages of this criterion can be summarized in two main points: the criterion uses both stress and strain to predict failure and only one set of data is required (and used as the reference set) for calibration of the model parameters. On the other hand, the model is only applicable to unidirectional laminates. Predictions of the model for glass/epoxy and carbon/epoxy unidirectional laminates were in good agreement with experimental fatigue data in [22].

6.3 Experimental Results from Literature

The described fatigue theories were developed based on experimental data concerning unidirectional and multidirectional composite laminates. Selected data from the literature will be presented in the following paragraphs the demonstration of the application of the fatigue theories and the evaluation of their predictive ability.

All available experimental fatigue data were treated in the same manner. S–N curves of the form:

$$\sigma_a \text{ or } \sigma_{\max} = \sigma_o N^{-1/k} \qquad (6.28)$$

were fitted to the available databases and used in calculations of fatigue life predictions.

6.3.1 Uniaxial Loading

6.3.1.1 Glass/Epoxy, UD, Flat Specimens (Hashin and Rotem)

A complete set of off-axis fatigue results from unidirectional E-glass/epoxy specimens are presented in [8]. The experiments were performed at three different fatigue frequencies, 34, 19 and 1.8 Hz, and were concluded when the specimen was fractured or when the limit of 10^6 cycles was reached. Tension-tension fatigue loading under the R-ratio of 0.1 was applied. The experimental results for specimens at $\theta = 5°, 10°, 15°, 20°, 30°$ and $60°$ are presented in Figs. 6.1 and 6.2. The parameters for the S–N equations, according to Eq. 6.28, are given in Table 6.1. The S–N curves along the transverse direction (90°) and at 10° were used for calculation of the S–N curve under shear as proposed by the authors [8], using Eq. 6.3.

Fig. 6.1 Experimental
results Hashin–Rotem [8].
5°, 10°, 15°, off-axis

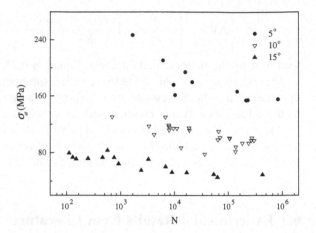

Fig. 6.2 Experimental
results, Hashin–Rotem [8].
20°, 30°, 60°, off-axis

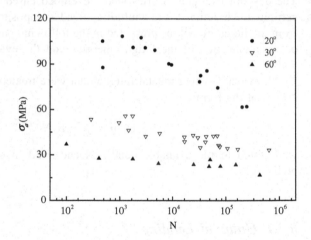

6.3.1.2 Graphite/Epoxy, UD, Flat Specimens (Awerbuch and Hahn)

Awerbuch and Hahn [24] presented the results of an experimental program
comprising static and fatigue tests on unidirectional carbon/epoxy specimens.

Constant amplitude loading under $R = 0.1$ at a frequency of 18 Hz was applied
until failure or the limit of 10^6 fatigue cycles. The maximum applied cyclic stress
was between 50 and 70% of the static strength of the material along each off-axis
angle. The experimental results for several on- and off-axis angles are presented in
Table 6.2. The off-axis static strengths were calculated using the Tsai-Wu [7]
failure criterion and the available static strength data along the principal directions
of the examined material. The experimental results from specimens cut at different
off-axis angles between 10° and 90° are given in Figs. 6.3 and 6.4.

Table 6.1 S–N curves and corresponding static strengths for experimental data of Hashin–Rotem [8]

Off-axis angle θ	UTS (MPa)	S–N curve (MPa)
0°	1236.09	$\sigma_a = 1740.00N^{-0.0786}$
5°	434.26*	$\sigma_a = 478.65N^{-0.0947}$
10°	216.08*	$\sigma_a = 354.49N^{-0.1186}$
15°	142.98*	$\sigma_a = 221.25N^{-0.1022}$
20°	106.24*	$\sigma_a = 125.24N^{-0.0880}$
30°	69.45*	$\sigma_a = 101.41N^{-0.0913}$
60°	34.81*	$\sigma_a = 56.26N^{-0.0864}$
90°	28.45	$\sigma_a = 41.04N^{-0.0860}$
Shear	37.96*	$\sigma_a = 64.35N^{-0.1118}*$

*value calculated using Eq. 6.3

Table 6.2 S–N curves and corresponding static strengths for experimental data of Awerbuch and Hahn [24]

Off-axis angle θ	UTS (MPa)	UCS (MPa)	S–N curve (MPa)
0°	1836.0	1836.0	$\sigma_{max} = 1043.29N^{-0.0458}$
10°	458.0		$\sigma_{max} = 572.77N^{-0.0893}$
20°	211.0		$\sigma_{max} = 263.75N^{-0.07746}$
30°	129.0		
45°	94.0		$\sigma_{max} = 168.60N^{-0.1250}$
60°	70.0		$\sigma_{max} = 80.18N^{-0.0533}$
90°	56.9	207.0	$\sigma_{max} = 93.61N^{-0.1103}$
Shear	93.0	93.0	

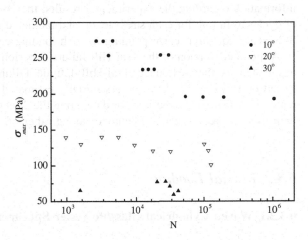

Fig. 6.3 Experimental results, Awerbuch-Hahn [24]. 10°, 20°, 30°, off-axis

As shown in Fig. 6.4, the quality of the off-axis data at high angles is very low. Therefore, only the data from specimens cut at 10°, 20° and 45° off-axis will be used in the next section for evaluation of the fatigue failure criteria.

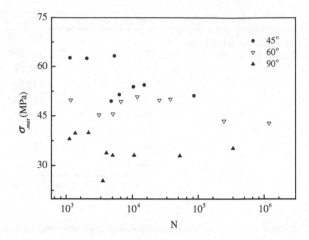

Fig. 6.4 Experimental results, Awerbuch-Hahn [24]. 45°, 60°, 90°, off-axis

6.3.1.3 AS4 Carbon/PEEK APC2, Flat Specimens (Jen-Lee)

The experimental program of Jen and Lee [13, 14] consisted of quasi-static and constant amplitude fatigue tests on unidirectional, cross-ply and pseudo-isotropic laminates. Different R-ratios were selected for the fatigue loading, resulting in different applied loading patterns covering both tension-tension and compression-compression fatigue. A constant frequency of 5 Hz was maintained for all studied cases and it was found that Log-Log S–N curves could fit the experimental data accurately. The derived S–N equations for all examined unidirectional specimens are given in Table 6.3. It should be noted that the authors do not provide analytical information regarding the experimental results; they only give the average values of cycles to failure for each stress level, and tabulated information concerning the S–N curve equations corresponding to each loading case.

A modified version of the Tsai-Hill failure criterion proposed by the authors is designated as the "extended Tsai-Hill fatigue failure criterion" (ET-H FFC). Based on this criterion, the authors initially estimated the shear strength of their material by using the experimental data along the directions of 0°, 90° and 45° and then tried to predict material behavior along other off-axis directions.

6.3.2 Biaxial Loading

6.3.2.1 Woven Cylindrical Glass/Polyester Specimens (Owen-Griffiths)

Constant amplitude fatigue data from an experimental program comprising biaxial tests on cylindrical woven, glass/polyester specimens is presented in Owen and Griffiths [10]. An axial load together with internal pressure was applied on the specimens in order to produce the desired biaxial stress states.

Table 6.3 S–N curves for experimental data of Jen and Lee [13, 14]

θ	UTS (MPa)	UCS (MPa)	S–N curve (MPa)			
			$R = 0$	$R = 0.2$	$R = -\infty$	$R = 5$
0°	2 128	955	$\sigma_{max} = 2673.00N^{-0.0503}$	$\sigma_{max} = 2604.47N^{-0.0446}$	$\sigma_{max} = 1235.55N^{-0.0620}$	$\sigma_{max} = 1192.50N^{-0.0548}$
15°	459	461	$\sigma_{max} = 603.67N^{-0.0680}$	$\sigma_{max} = 589.13N^{-0.0630}$	$\sigma_{max} = 608.92N^{-0.0675}$	$\sigma_{max} = 648.78N^{-0.0707}$
30°	221	275	$\sigma_{max} = 287.36N^{-0.0783}$		$\sigma_{max} = 385.23N^{-0.0809}$	
45°	151	228	$\sigma_{max} = 175.54N^{-0.0710}$	$\sigma_{max} = 173.44N^{-0.0622}$	$\sigma_{max} = 277.57N^{-0.0780}$	$\sigma_{max} = 366.51N^{-0.0952}$
60°	116	215	$\sigma_{max} = 158.49N^{-0.0910}$	$\sigma_{max} = 153.70N^{-0.0804}$	$\sigma_{max} = 242.05N^{-0.0677}$	$\sigma_{max} = 338.66N^{-0.0938}$
75°	96	211	$\sigma_{max} = 114.46N^{-0.0782}$		$\sigma_{max} = 206.46N^{-0.0658}$	
90°	93	206	$\sigma_{max} = 116.81N^{-0.0813}$	$\sigma_{max} = 126.80N^{-0.0774}$	$\sigma_{max} = 262.68N^{-0.0879}$	$\sigma_{max} = 274.45N^{-0.0851}$
Shear	133	133				

Fig. 6.5 Experimental results, Owen-Griffiths [10]. 0°, 45°

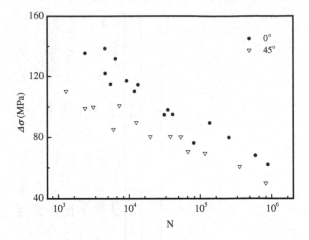

Table 6.4 S–N curves for experimental data of Owen and Griffiths [10]

Angle θ	S–N curve (Mpa)
0°	$\Delta\sigma = 416.75N^{-0.1390}$
45°	$\Delta\sigma = 265.44N^{-0.1183}$

Constant amplitude cyclic loads under a stress ratio, $R = 0$, were applied on the specimens at a constant frequency of 1.8 Hz.

The woven material used exhibited the same properties along the longitudinal and transverse directions. Therefore, only two S–N curves are required (one along the longitudinal/transverse directions and one for shear) for application of the examined fatigue failure criteria.

The authors presented uniaxial experimental data for on-axis and 45° off-axis specimens as shown in Fig. 6.5 as number of cycles to failure under specific values of the stress range. The fitted Log-Log S–N equations are given in Table 6.4. They also performed experiments under five biaxial stress ratios (axial stress over hoop stress) ($a = \sigma_{ax}/\sigma_{hp}$) between –1 and 1 (–1, –0.5, 0, 0.5, 1) in order to simulate the failure loci for different numbers of cycles. The biaxial experimental results are presented in Figs. 6.6 and 6.7 and tabulated in Tables 6.5 and 6.6.

6.3.2.2 Woven Cylindrical Glass/Polyester Specimens (Fujii-Lin)

Another biaxial program was carried out by Fujii and Lin [11]. Cylindrical woven, glass/polyester specimens were tested under biaxial tensile-torsional loading under constant amplitude conditions. A biaxial stress field containing normal and shear stress components developed as a result. The frequency was limited to 2 Hz in order to avoid the development of high temperatures during the application of the loads under an $R = 0$. The uniaxial test results

Fig. 6.6 Experimental
results, Owen-Griffiths [10].
Failure loci, specimens at 0°

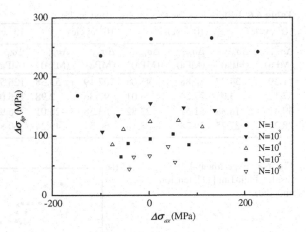

Fig. 6.7 Experimental
results Owen-Griffiths [10].
Failure loci, specimens at 45°

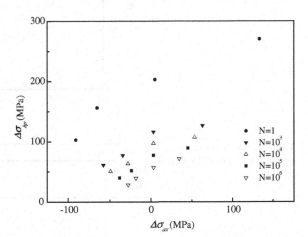

Table 6.5 Failure loci for specimens tested at 0°

10^6 cycles		10^5 cycles		10^4 cycles		10^3 cycles		1 cycle	
$\Delta\sigma_{hp}$ (MPa)	$\Delta\sigma_{ax}$ (MPa)	$\Delta\sigma_{hp}$ (MPa)	$\Delta\sigma_{ax}$ (MPa)	$\Delta\sigma_{hp}$ (MPa)	$\Delta\sigma_{ax}$ (MPa)	$\Delta\sigma_{hp}$ (MPa)	$\Delta\sigma_{ax}$ (MPa)	$\Delta\sigma_{hp}$ (MPa)	$\Delta\sigma_{ax}$ (MPa)
55.01	53.67	84.23	82.67	114.49	110.64	142.10	136.96	241.58	229.11
80.92	39.21	102.48	49.74	125.25	60.68	146.60	71.82	265.13	130.42
65.98	1.05	94.49	2.38	124.02	2.07	154.37	2.79	263.93	4.19
64.06	−31.74	86.91	−43.55	110.77	−53.10	134.02	−63.676	235.57	−100.01
43.93	−41.24	64.56	−58.38	85.19	−76.14	106.44	−96.97	167.71	−146.76

Table 6.6 Failure loci for specimens tested at 45°

10^6 cycles		10^5 cycles		10^4 cycles		10^3 cycles		1 cycle	
$\Delta\sigma_{hp}$ (MPa)	$\Delta\sigma_{ax}$ (MPa)	$\Delta\sigma_{hp}$ (MPa)	$\Delta\sigma_{ax}$ (MPa)	$\Delta\sigma_{hp}$ (MPa)	$\Delta\sigma_{ax}$ (MPa)	$\Delta\sigma_{hp}$ (MPa)	$\Delta\sigma_{ax}$ (MPa)	$\Delta\sigma_{hp}$ (MPa)	$\Delta\sigma_{ax}$ (MPa)
71.39	34.27	88.88	45.08	107.19	53.38	126.51	62.94	270.31	133.98
57.12	3.05	77.24	3.01	97.36	2.98	116.06	2.95	203.26	4.70
39.41	−18.39	51.59	−23.82	63.38	−28.01	76.98	−34.28	156.21	−65.02
28.21	−27.75	39.78	−38.18	50.54	−49.24	60.89	−57.80	102.72	−90.57

Fig. 6.8 Experimental results, Fujii-Lin [11], tension and shear

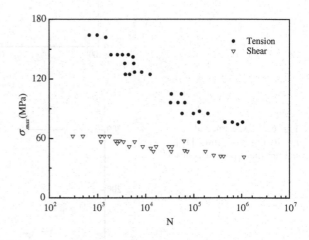

Table 6.7 S–N curves for experimental data of Fujii and Lin [11]

Loading pattern	S–N curve (MPa)
Tension	$\sigma_{max} = 405.65N^{-0.1309}$
Shear	$\sigma_{max} = 97.05N^{-0.0649}$

Fig. 6.9 Experimental results, Fujii-Lin [11], failure loci

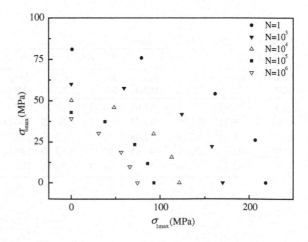

Table 6.8 Failure loci for specimens tested at 45°

1 cycle		10^3 cycles		10^4 cycles		10^5 cycles		10^6 cycles	
σ_{1max} (MPa)	σ_{6max} (MPa)	σ_{1max} (MPa)	σ_{6max} (MPa)	σ_{1max} (MPa)	σ_{6max} (MPa)	σ_{1max} (MPa)	σ_{6max} (MPa)	σ_{1max} (MPa)	σ_{6max} (MPa)
1.22	81.05	0.23	60.05	0.42	50.25	0.16	42.91	0.30	39.17
79.14	75.89	59.52	57.63	48.29	45.99	37.96	37.31	30.86	30.29
161.77	54.20	124.40	41.76	92.85	29.90	71.25	23.38	56.32	18.74
206.97	26.13	158.03	22.28	112.95	15.69	85.89	11.79	65.77	9.77
218.74	0.00	169.92	0.00	121.4	0.00	93.12	0.00	74.39	0.00

(with biaxial stress ratios $a = \sigma_1/\sigma_6 = 1/0$, tension and $a = 0/1$ shear) are presented in Fig. 6.8, while the fitted Log-Log equations are given in Table 6.7.

Three biaxial stress ratios were obtained in addition to the uniaxial loading cases described earlier. 7/1, 3/1, 1/1, and the experimental results are presented in Fig. 6.9 and tabulated in Table 6.8.

6.4 Fatigue Life Prediction

The fatigue failure theories described under Sect. 6.2 will be used here to predict the fatigue behavior of the composite material systems presented in Chap. 2 and Sect. 6.3. It should be mentioned however that some of the aforementioned criteria, e.g., that proposed by Kawai [18], or that of Shokrieh-Taheri [22], are limited to unidirectional materials and therefore not able to predict the off-axis behavior of the multidirectional composite laminate in Chap. 2, nor the behavior of the woven composites tested under biaxial loading and described in Sect. 6.3. In addition to these, the Hashin-Rotem [8] criterion is not applicable for either multidirectional or woven materials since it is not possible to define whether the dominant failure mode is the fiber or matrix mode.

The Log-Log S–N curve type, e.g., Eq. 6.28, was used in all cases for fitting of the fatigue data for consistency, although the Lin-Log representation was used in the original versions of some of the criteria, e.g., FWE.

6.4.1 Uniaxial Loading

6.4.1.1 Glass/Epoxy, UD, Flat Specimens (Hashin and Rotem)

All examined failure theories were applicable for the estimation of the off-axis fatigue behavior of the material system presented in the paper of Hashin and Rotem [8]. Fatigue functions (S–N curves) of the Log-Log form were used for all theories, when necessary. The comparison of the predictions is presented in

Fig. 6.10 Predicted S–N curves for 5°, 10° and 20° off-axis specimens [8]. FTPF, HR and FWE criteria

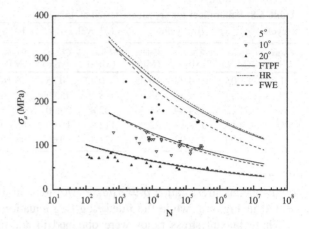

Fig. 6.11 Predicted S–N curves for 15°, 30° and 60° off-axis specimens [8], FTPF, HR and FWE criteria

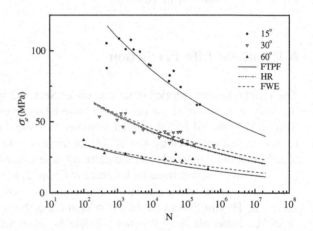

Figs. 6.10, 6.11, 6.12 and 6.13 for the fatigue data of specimens cut along the off-axis directions between 5° and 60°.

The S–N curve at 15° off-axis was used as the reference curve for the FWE, KW and ST criteria. As shown in Figs. 6.10, 6.11, 6.12 and 6.13, all fatigue theories, except KW, overestimate the fatigue behavior for small off-axis angles (5° and 10°), while they are accurate for the off-axis directions between 15° and 60°. In contrast, KW is more accurate for 5°, 10° and 20°, while it tends to underestimate the material behavior for bigger off-axis angles, e.g., 30° and 60°, see Fig. 6.13. The equations for the predicted S–N curves are given in Table 6.9.

The curves predicted using the HR criterion are the most accurate of the set of plotted curves with $R^2 > 0.86$. FTPF and SB predictions exhibited similar accuracy ($R^2 > 0.82$) although they tend to overestimate the life at the low cycle fatigue regime and be more conservative at the high cycle fatigue regime. FWE, ST and KW produced less accurate predictions.

Fig. 6.12 Predicted S–N
curves for 5°, 10° and 20°
off-axis specimens [8], SB,
KW and ST criteria

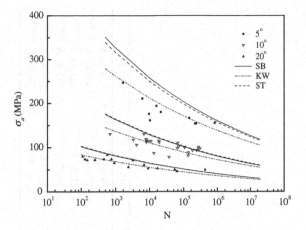

Fig. 6.13 Predicted S–N
curves for 15°, 30° and 60°
off-axis specimens [8], SB,
KW and ST criteria

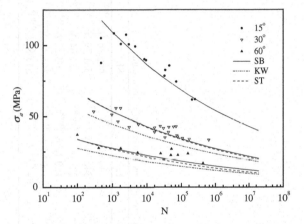

FTPF and SB predictions are similar since for this case, where the compressive strengths are considered equal to the tensile strengths, the only difference between the Tsai-Hill and Tsai-Hahn criteria, on which the two fatigue theories were founded, is the interaction term ($1/X^2$ for Tsai-Hill and $1/XY$ for Tsai-Hahn), which does not greatly affect the predictions compared with the rest of the terms of the failure tensor polynomial.

The 15° off-axis S–N curve has been arbitrarily selected as the reference one for the application of the FWE, KW and ST criteria. However, as also reported elsewhere, e.g., [16, 20, 30], criteria based on this "master" curve concept (like FWE, KW and ST) are quite sensitive to the selection of the reference curve. A comparison of the theoretical predictions of the FWE criterion to the 30° off-axis fatigue data, using different reference curves, is presented in Fig. 6.14. As shown, selection of the S–N curve at 15° off-axis seems to produce the most accurate theoretical results, while the rest of the estimated S–N curves are conservative (ref. 5° and 20°) or highly non-conservative, especially for the low-cycle fatigue region (ref. 60°).

Table 6.9 Analytical comparison of fatigue failure theory predictions with experimental data

Off-axis angle	FTPF, σ_a	HR, σ_a	FEW, σ_a	SB, σ_a	KW, σ_a	ST, σ_a
5°	$639.96\,N^{-0.1019}$	$666.16\,N^{-0.1031}$	$740.96\,N^{-0.1251}$	$666.48\,N^{-0.1031}$	$459.93\,N^{-0.0921}$	$638.36\,N^{-0.1017}$
10°	$332.95\,N^{-0.1027}$	$331.15\,N^{-0.1026}$	$346.74\,N^{-0.1107}$	$331.29\,N^{-0.1027}$	$257.66\,N^{-0.0921}$	$331.65\,N^{-0.1017}$
15°	$221.64\,N^{-0.1022}$	$221.64\,N^{-0.1022}$	$221.64\,N^{-0.1022}$	$221.64\,N^{-0.1022}$	$221.31\,N^{-0.0921}$	$221.31\,N^{-0.1017}$
20°	$164.54\,N^{-0.1011}$	$162.14\,N^{-0.1008}$	$160.16\,N^{-0.0961}$	$162.22\,N^{-0.1008}$	$128.12\,N^{-0.0921}$	$164.93\,N^{-0.1017}$
30°	$106.58\,N^{-0.0980}$	$105.23\,N^{-0.0981}$	$100.85\,N^{-0.0873}$	$105.26\,N^{-0.0981}$	$83.91\,N^{-0.0921}$	$108.02\,N^{-0.1017}$
60°	$51.25\,N^{-0.0896}$	$51.07\,N^{-0.0896}$	$47.57\,N^{-0.0731}$	$51.08\,N^{-0.0896}$	$42.09\,N^{-0.0921}$	$54.20\,N^{-0.1017}$

Fig. 6.14 Sensitivity of FWE criterion to reference curve

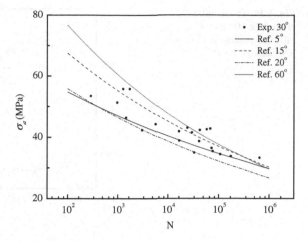

Fig. 6.15 Predicted S–N curves for 10° and 45° off-axis specimens [23], FTPF, HR and FWE criteria

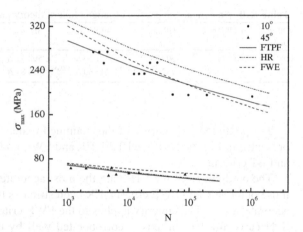

6.4.1.2 Graphite/Epoxy, UD, Flat Specimens (Awerbuch and Hahn)

The results of the application of all the examined theories on the fatigue data of the material presented by Awerbuch and Hahn in [24] are presented in Figs. 6.15 and 6.16. The S–N curve at 20° off-axis was used as the reference curve for all the examined fatigue theories, when necessary, and the S–N curves at 10° and 45° were used for the evaluation of their predictive accuracy. The fatigue shear strength necessary for application of the FTPF and SB criteria was estimated by fitting the 20° off-axis fatigue data to Eq. 6.17, while the same dataset (the 20° off-axis S–N plus the S–N under the transverse direction) was used together with the corresponding static strength data, using Eq. 6.3 for estimation of the fatigue shear strength based on the HR criterion.

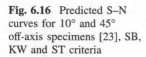

Fig. 6.16 Predicted S–N curves for 10° and 45° off-axis specimens [23], SB, KW and ST criteria

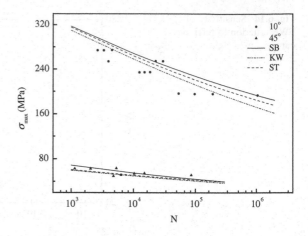

Table 6.10 Analytical comparison of fatigue failure theory predictions

Off-axis angle	FTPF, σ_{max}	HR, σ_{max}	FEW, σ_{max}	SB, σ_{max}	KW, σ_{max}	ST, σ_{max}
10°	$464.74\ N^{-0.0674}$	$509.98\ N^{-0.0649}$	$591.30\ N^{-0.0887}$	$508.53\ N^{-0.0700}$	$608.98\ N^{-0.0917}$	$538.60\ N^{-0.0775}$
45°	$136.66\ N^{-0.0984}$	$142.93\ N^{-0.1004}$	$113.60\ N^{-0.0658}$	$136.84\ N^{-0.0985}$	$117.36\ N^{-0.0917}$	$103.88\ N^{-0.0775}$

The predicted S–N curves of the examined material at 10° and 45° off-axis are presented in Fig. 6.15 for the FTPF, HR and FWE and in Fig. 6.16 for the SB, KW and ST criteria.

The predictions of the FTPF are the most accurate and consistent compared to those of the HR and FWE criteria. HR overestimates the behavior at 10° while it is accurate at 45°. The contrary applies to the FWE criterion for which the predicted S–N curve for 10° off-axis is corroborated well by the experimental data, being steeper however than the curve predicted from the FTPF. For the 45° off-axis specimens, the FWE criterion gives an optimistic prediction of the fatigue life.

In general, the predictions given by the examined theories are corroborated well by the experimental data, for both the small off-axis angle, 10°, and the 45°. However, KW and ST seem to be more conservative than SB for both the examined cases.

The estimated S–N curves in the form of the Log-Log curve are given in Table 6.10.

6.4.1.3 AS4 Carbon/PEEK APC2, Flat Specimens (Jen-Lee)

The material data from [13 and 14] were used for the evaluation of the predictive ability of the examined fatigue theories and detailed investigation of the influence of the reference curve selection on the accuracy of the fatigue theories that follow

Fig. 6.17 Predicted S–N curves for 15° and 60° off-axis specimens under $R = 5$, FTPF, HR and FWE criteria. Ref. 45°

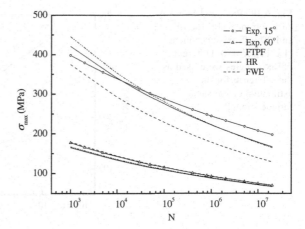

Fig. 6.18 Predicted S–N curves for 15° and 60° off-axis specimens under $R = 5$, SB, KW and ST criteria. Ref. 45°

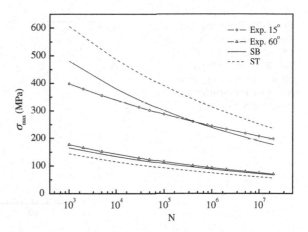

the "master curve" concept, e.g., FWE, KW and ST. The FWE criterion will be used for the demonstration. The abundance of experimental data in the database of Jen and Lee [13, 14] allows the examination of the sensitivity of the polynomial-based fatigue failure criteria predictions on the off-axis angle also used for derivation of the material's shear fatigue strength. The FTPF criterion, which was proved more accurate than HR and SB for the previously examined fatigue datasets, will be used as a representative of this type of theory for the comparisons to the "master curve"-based theories.

The theoretical predictions are presented in Figs. 6.17 and 6.18 for the loading case of compression-compression under $R = 5$. The KW criterion cannot be applied in this case since non-negative stresses are present. The S–N curve at 45° was selected as the common reference for all the employed criteria for consistency. The experimental fatigue data at 15° and 60° off-axis angles were used for the comparisons. As presented in Figs. 6.17 and 6.18, the polynomial-based fatigue

Fig. 6.19 Predicted S–N curves from FTPF for 15°, and 30° off-axis specimens [13, 14], $R = 0$. Different off-axis angles were used as the reference for the estimation of the shear fatigue strength

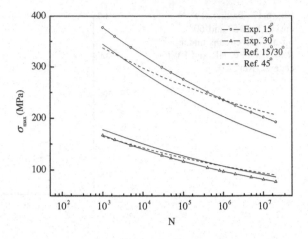

Fig. 6.20 Predicted S–N curves from FTPF for 30° and 75° off-axis specimens [13, 14], $R = -\infty$. Different off-axis angles were used as reference for estimation of shear fatigue strength

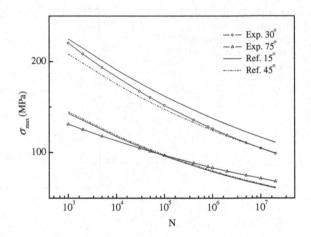

failure theories produce more accurate theoretical predictions than those based on the "master curve" concept. The FWE criterion is conservative, as presented in Fig. 6.17, while the ST criterion is conservative for the off-axis angle of 60°, but becomes highly non-conservative for the off-axis angle of 15°, see Fig. 6.18.

An assessment of the sensitivity of the examined fatigue theories has also been made in this section. FTPF and FWE were selected and applied for estimation of the off-axis fatigue life of the examined material, using different reference curves. The theoretical predictions of FTPF for specimens cut at the off-axis angles of 15° and 30° and tested under tension-tension loading, at $R = 0$, are presented in Fig. 6.19, together with the fitted curve to the experimental data. Two predicted curves are shown for each case, corresponding to two different reference curves. A similar process was followed ($R = -\infty$) and different off-axis angles (30° and 75°) and the results are presented in Fig. 6.20.

Fig. 6.21 Predicted S–N curves from FWE for 15° and 30° off-axis specimens [13, 14], $R = 0$. Different off-axis angles were used as reference for estimation of shear fatigue strength

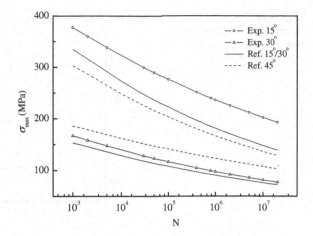

Fig. 6.22 Predicted S–N curves from FWE for 30° and 75° off-axis specimens [13, 14], $R = -\infty$. Different off-axis angles were used as reference for estimation of shear fatigue strength

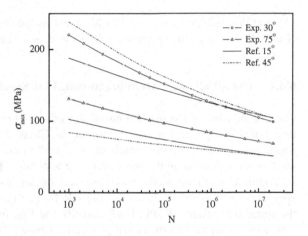

Observation of Figs. 6.19 and 6.20 leads to the conclusion that the FTPF fatigue theory is relatively insensitive to the selection of the reference curve used for derivation of the shear fatigue strength of the material by means of Eq. 6.18 and is able to simulate the fatigue behavior of the examined material with acceptable accuracy. All the estimated S–N curves lie on a narrow band on the S–N plane with differences between them, and between them and the experimental data, of less than one decade of life, which is in a lot of cases the typical scatter of the experimental results for composite laminates.

The same process was followed for the FWE criterion and the results are presented in Figs. 6.21 and 6.22. FWE is shown to be quite sensitive to the selection of the reference curve as it can produce accurate, conservative or non-conservative theoretical predictions for different choices. Independent of the accuracy of the theoretical predictions of the FWE criterion, the presented scatter is very large, in some of the examined cases reaching even three to four decades of

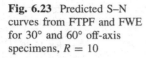

Fig. 6.23 Predicted S–N curves from FTPF and FWE for 30° and 60° off-axis specimens, $R = 10$

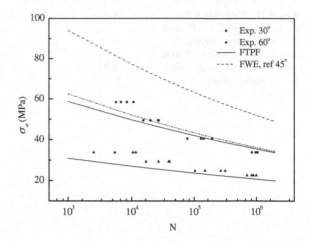

life (see Fig. 6.22, e.g., $\sigma_{\max} = 100$ MPa, estimated N equals ca. 10^3, when the S–N at 15° is used as the reference, while the actual value is ca. 10^5).

6.4.1.4 Glass/Polyester, Multidirectional, Flat Specimens (Chap. 2)

The comparison of the predictions of the off-axis fatigue behavior of the multi-directional composite laminate presented in Chap. 2 is made in this paragraph. Since this material is not unidirectional and moreover negative mean cyclic stresses were applied on the specimens, the KW and ST criteria cannot be applied. In addition, a mixed mode was observed during the failure of the examined specimens and therefore the use of the HR fatigue theory is also excluded. From the remaining criteria–FTPF, FWE and SB–the first two will be employed in the following as representative of the polynomial-based theories (FTPF) and theories based on the master curve concept (FWE).

Calculations based on the two examined fatigue theories were performed using the material data presented in Chap. 2. Estimation of the fatigue life of specimens cut at different off-axis angles and loaded under different stress ratios was possible. For the application of the FTPF criterion on this dataset, it had previously been proved [17] that the derivation of the shear fatigue strength based on the experimentally derived curve for 45° off-axis was appropriate. The shear fatigue strength for S–N formulations were derived by dividing the corresponding S–N curves at 45° by a factor of 2.2 [17]. The S–N curves at 45° were also used as the reference curves for the application of the FWE criterion.

The results, presented in Figs. 6.23, 6.24 and 6.25, indicate a superior predictive ability of the FTPF criterion compared to the FWE. In all cases but one ($R = 0.1$, 15° off-axis, Fig. 6.25) FTPF's predictions are accurate and on the safe side, i.e., the predicted S–N curves are conservative when compared to the experimental data. On the other hand, the predictions produced by application of

Fig. 6.24 Predicted S–N curves from FTPF and FWE for 30° and 60° off-axis specimens, $R = -1$

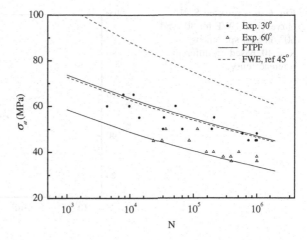

Fig. 6.25 Predicted S–N curves from FTPF and FWE for 15° and 75° off-axis specimens, $R = 0.1$

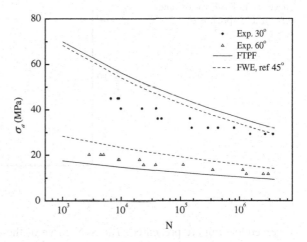

the FWE criterion are mainly non-conservative, estimating a longer fatigue life than that experimentally determined for all the examined cases. The difference between the theoretical and actual lifetime ranges between half (as it is also the average scatter of the available experimental data) and three decades of life, as for example presented in Fig. 6.23 for the 60° off-axis angle.

It should be mentioned however that FWE criterion can be more accurate if an S–N along a different on- or off-axis angle will be employed as the reference curve. The selection, for example, of the on-axis S–N curve as the reference for the loading case of $R = 10$ is proved more appropriate than the 45° off-axis, as presented in Fig. 6.26, where the predictions of the FWE criterion of the 30° and 60° off-axis angles are compared to the experimental data. The examined fatigue theory can even be more accurate, see for example Fig. 6.27, where the predictions of the fatigue life of the on-axis and 60° off-axis specimens fatigued under

Fig. 6.26 Predicted S–N
curves from FWE for 30° and
60° off-axis specimens,
$R = 10$, based on different
reference curves

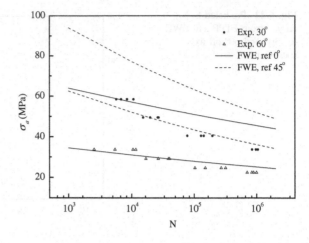

Fig. 6.27 Predicted S–N
curves from FWE for 0°, and
60° off-axis specimens,
$R = -1$, ref. 30°

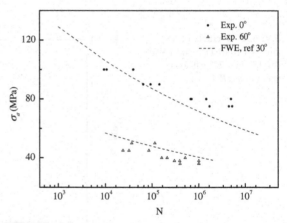

reversed loading are presented. The S–N curve of the specimen cut at 30° off-axis
has been used as the reference for the application of the FWE criterion and very
accurate results were obtained. Although these very good predictions strengthened
the validity of the FWE criterion, it also proved its main disadvantage: it is very
sensitive to the selection of the reference angle and its results can be accurate or
not according to a successful or unsuccessful decision concerning the selection of
the reference curve.

6.4.2 Biaxial Loading

Biaxial fatigue data from the literature will also be employed for evaluation of the
predictive ability of the aforementioned fatigue theories. The available biaxial
fatigue data refer to woven cylindrical specimens either loaded under biaxial plane

Fig. 6.28 Comparison of theoretical (predicted by FTPF, SB and FWE) failure loci vs. experimental data of cylindrical specimens loaded at 0°. FWE, ref. 0°

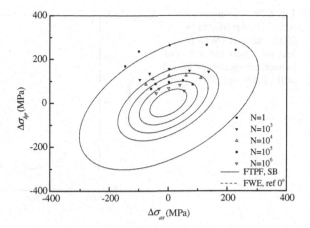

stress state comprising axial and tangential normal stresses, Owen and Griffiths [10], or biaxial plane stress states comprising axial normal and torsional (shear) stress components as presented by Fujii and Lin [11].

Only three of the examined fatigue theories are applicable to the material used in these studies, the FTPF, SB and FWE, while the rest cannot be applied for different reasons. HR is not applicable since for the woven material it is not possible to separate the failure modes into fiber- or matrix-dominated. On the other hand, KW and ST are applicable only for unidirectional composite laminates.

6.4.2.1 Biaxial Loading of Thin-Walled Tubes

Owen and Griffiths [10] examined glass/polyester thin-walled tubes under combined axial loading and internal pressure under the stress ratio of $R = 0$. Additionally, laminates of the same material were tested at fiber directions of $0°$ and $45°$ to estimate the axial and shear fatigue strength of the examined material. The longitudinal and transverse static strengths were $X = Y = 262.2$ MPa, while shear strength was $S = 110$ MPa. The shear fatigue strength was estimated based on the FTPF criterion Eq. 6.18 using the S–N curve at 45°. Both S–N curves from the specimens loaded at 0° and 45° were used as reference for application of the FWE criterion.

The fatigue theories were applied and the predicted failure loci are shown in Figs. 6.28, 6.29, 6.30 and 6.31 for the cylindrical specimens of [10] loaded at 0° and 45°, respectively.

As presented in Figs. 6.28, 6.29, 6.30 and 6.31, the FTPF and SB coincide since the examined woven material exhibits equal strength in the longitudinal and transverse directions, and they generally produce acceptable predictions. The FWE criterion also leads to accurate theoretical predictions, similar to the predictions of the previous two criteria.

Fig. 6.29 Comparison of theoretical (predicted by FTPF, SB and FWE) failure loci vs. experimental data of cylindrical specimens loaded at 0°. FWE, ref. 45°

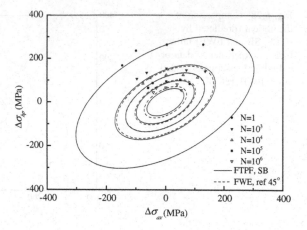

Fig. 6.30 Comparison of theoretical (predicted by FTPF, SB and FWE) failure loci vs. experimental data of cylindrical specimens loaded at 45°. FWE, ref. 0°

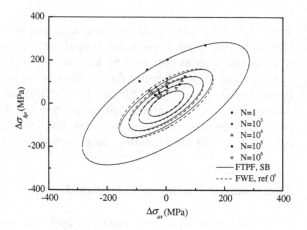

Fig. 6.31 Comparison of theoretical (predicted by FTPF, SB and FWE) failure loci vs. experimental data of cylindrical specimens loaded at 45°. FWE, ref. 45°

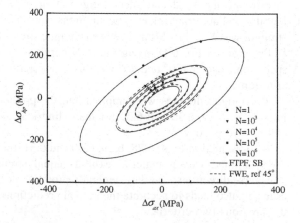

This occurs because all the examined theories result in similar relationships for the estimation of the failure loci for each loading condition. For the woven material of [10] loaded by an on-axis biaxial plane stress field comprising the normal stress components σ_1 and σ_2, the failure locus for each biaxial ratio, $a_1 = \sigma_1/\sigma_2$, can be estimated by:

$$\sigma_x^{FTPF} = \pm \frac{X(R,N,\theta)}{\sqrt{1+\alpha_1^2}-\alpha_1} \tag{6.29}$$

When the FTPF or SB criterion is used, while for the FWE the relationship becomes:

$$\sigma_x^{FWE} = \pm \frac{X}{X_r} \frac{X(R,N,\theta)}{\sqrt{1+\alpha_1^2}-\alpha_1} \tag{6.30}$$

with X denoting the static strength along the axial (longitudinal) direction, and $X(R, N, \theta)$ denoting the corresponding S–N curve, as a function of the fatigue stress ratio, R, number of cycles, N and angle, θ. The subscript r designates the reference angle.

According to Eqs. 6.29 and 6.30, when the on-axis direction is used as the reference, since the ratio $X/X_r = 1$, and $X(R, N, \theta) = X_r(R, N, \theta)$, the predictions of all criteria must coincide, as proved in Fig. 6.28. The differences between the FTPF or SB and FWE predictions seen in Fig. 6.29 are due to the use of the S–N curve of the 45° specimen. However, for the examined material the use of this reference has little influence on the results (the S–N curve of the on-axis specimens can accurately be predicted from the S–N at 45° and the FWE criterion). Therefore, the differences presented in Fig. 6.29 are not significant.

For the derivation of the failure loci at 45° for the same plane stress field, Eqs. 6.29 and 6.30 become:

$$\sigma_x^{FTPF} = \pm \frac{2X(R,N,\theta)}{\sqrt{(1+\alpha_1)^2+\left(\frac{X(R,N,\theta)}{S(R,N,\theta)}\right)^2(\alpha_1-1)^2}} \tag{6.31}$$

$$\sigma_x^{FWE} = \pm \frac{X}{X_r} \frac{2X(R,N,\theta)}{\sqrt{(1+\alpha_1)^2+\left(\frac{X}{S}\right)^2(\alpha_1-1)^2}} \tag{6.32}$$

with S and $S(R, N, \theta)$ denoting the shear static and fatigue strengths, respectively.

In this case, the similarity of the predictions, in addition to the effect of the reference curve selection, are also affected by the resemblance of the ratio between X and S when compared to the ratio $X(R, N, \theta)/S(R, N, \theta)$. The aforementioned ratios are given in Table 6.11 for different numbers of cycles between 1,000 and 1,000,000.

As shown in Table 6.11, the ratio between the axial strength and the shear strength (considered as the reference for application of the FWE criterion) is

Table 6.11 Calculated difference between static and fatigue strength ratios (X/S) of woven material presented in [10]

N	$A = X/S$ (262.2 MPa/ 110 MPa)	$B = X(R,N,\theta)/S(R,N,\theta)$ 416.74 $N^{-0.139}$/ 142.52 $N^{-0.1169}$	Difference % $(A-B/A)$ 100
1,000	2.38	2.51	5.47
10,000	2.38	2.39	0.23
100,000	2.38	2.28	−4.74
1,000,000	2.38	2.15	−9.47

Fig. 6.32 Comparison of theoretical (predicted by FTPF, SB and FWE) failure loci vs. experimental data of cylindrical specimens loaded at 0°. FWE, ref. 0°

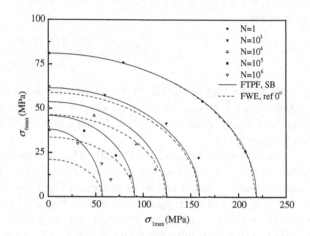

similar to the ratio between the corresponding fatigue strengths for different number of cycles. However, the static strength ratio is observed to be lower than the corresponding fatigue strength ratio for low number of cycles, $<10^5$, while this trend is reversed for number of cycles higher than 10^5. This behavior is also illustrated in Figs. 6.30 and 6.31. For $N = 10^5$ the predicted failure loci almost coincide, while they present differences for lower and higher numbers of fatigue cycles.

6.4.2.2 Biaxial Tensile-Shear Loading of Thin-Walled Tubes [11]

Woven glass/polyester tubes were examined by Fujii and Lin [11] under tensile and shear (torsion) loading under $R = 0$. The authors collected data under five biaxial ratios ($\alpha = \sigma_1/\sigma_6$), 1/0, 7/1, 3/1 1/1 and 0/1. The longitudinal and transverse normal strengths were $X = Y = 218.74$ MPa, while the shear strength was $S = 81.04$ MPa.

FTPF, SB and FWE were applicable for the examined case. As for the previously examined material, FTPF and SB predictions coincide. However, the FWE criterion proved very sensitive to the selection of the reference curve. As shown in Figs. 6.32 and 6.33, the criterion proved very conservative when the S–N curve

Fig. 6.33 Comparison of
theoretical (predicted by
FTPF, SB and FWE) failure
loci vs. experimental data of
cylindrical specimens loaded
at 0°. FWE, ref. shear

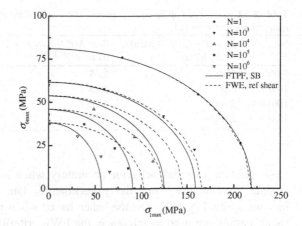

under the biaxial ratio 1/0 (tensile loading) is used as the reference, see Fig. 6.32,
but non-conservative when the S–N under shear loading was used as reference,
Fig. 6.33.

For the present plane stress state that consists of normal and shear stress
components, the failure loci for the two classes of examined criteria for any biaxial
stress ratio, $a_2 = \sigma_6/\sigma_1$ can be derived by the following equations:

$$\sigma_{xa_2}^{FTPF} = \pm \frac{X(R,N,\theta)}{\sqrt{1 + \left(\frac{X(R,N,\theta)}{S(R,N,\theta)}\right)^2 \alpha_2^2}} \tag{6.33}$$

for the FTPF and SB criteria, and:

$$\sigma_{xa_2}^{FWE} = \pm \frac{X}{S} \frac{S(R,N,\theta)}{\sqrt{1 + \left(\frac{X}{S}\right)^2 \alpha_2^2}} \tag{6.34}$$

when the shear S–N curve is used as reference, while the corresponding equation
for the reference based on the on-axis direction data becomes:

$$\sigma_{xa_2}^{FWE} = \pm \frac{X(R,N,\theta)}{\sqrt{1 + \left(\frac{X}{S}\right)^2 \alpha_2^2}} \tag{6.35}$$

The differences between the ratio of the static strengths and the corresponding
ratio of the fatigue strengths are tabulated in Table 6.12 for different number of
cycles between 1,000 and 1,000,000.

As shown, the difference between the two ratios increases with the number of
cycles. Therefore, as also observed in Figs. 6.32 and 6.33, the predictions of the
three criteria are similar for 1,000 cycles, while they become significantly different
as the fatigue life is increased, with FWE criterion being less accurate for both
cases of the selected reference angle. When the on-axis data is used as reference,

Table 6.12 Calculated difference between static and fatigue strength ratios (X/S) of woven material presented in [11]

N	A = X/S (218.74 MPa/ 81.32 MPa)	B = $X(R,N,\theta)/S(R,N,\theta)$ 405.65 $N^{-0.1309}$/ 97.05 $N^{-0.0649}$	Difference % (A−B/A) 100
1,000	2.69	2.65	−1.50
10,000	2.69	2.28	−15.39
100,000	2.69	1.96	−27.32
1,000,000	2.69	1.68	−37.57

FWE predicts material behavior accurately when $a_2 = 0$, while the predictions deviate significantly from the experimental data as the biaxial stress ratio increases, see Fig. 6.32. On the other hand, when the experimental data under shear loading are used as reference, the FWE criterion predicts well the behavior of the material for $a_2 = 1$, but deviates significantly from the actual material behavior for all the other cases. This observation confirms the aforementioned comments regarding the sensitivity of the FWE criterion for the selection of the reference data.

6.5 Evaluation of the Fatigue Theories

The following criteria were considered in order to evaluate the applicability of the examined fatigue theories and assess their influence on fatigue life prediction for the examined composite materials:

- Accuracy of predictions: quantified by the accuracy of predicting new S–N curves for off-axis angles or under biaxial loading patterns.
- Need for experimental data: quantified by the number of S–N curves required for the derivation of each model's parameters.
- Implemented assumptions and sensitivity to the data selected for modeling in relation to the consistency of the predictions: qualitative criterion.

   The comparison of the results for unidirectional materials [8, 13, 14, 24] shows that the polynomial fatigue failure criteria give similar predictions with reasonable accuracy. On the other hand, predictions involving criteria based on the use of a "master curve" proved to be very dependent on the selection of the appropriate reference curve. It was proved that in most of the examined cases these criteria were significantly conservative or non-conservative compared to existing experimental results.

   Only three of the examined fatigue theories (FTPF, SB and FWE) were applicable for the estimation of the fatigue life of the woven material tested under biaxial fatigue loading patterns. The results showed that the FTPF and SB criteria can be very accurate in this case, while the accuracy of the FWE criterion depends on the material. If the material off-axis fatigue behavior correlates well with the corresponding static behavior (e.g. $X/S \approx X(R, N, \theta)/S(R, N, \theta)$), life predictions can be accurate.

Regarding the need for experimental data for estimation of the model parameters, it is obvious that theories based on a reference curve and static strength data are superior to the other examined criteria. However these theories comprise a number of assumptions, e.g. combining of static and fatigue data. The use of quasi-static strength data for the derivation of fatigue curves (such as fatigue data for 1 or ¼ cycle) is also arguable. No complete study on this subject has been made. Previous publications, e.g. [31], showed that quasi-static data should not be a part of the S–N curve, especially when they have been acquired under strain rates much lower that those used in fatigue loading. The use of quasi-static data in the regression leads to incorrect slopes of the S–N curves in [31]. On the other hand, excluding quasi-static data improved the description of the fatigue data, but introduced errors in lifetime predictions when the low cycle regime is important, as for example for loading spectra with a few high-load cycles. Moreover, they are liable to be inconsistent since they cannot guarantee that the prediction will not be affected by the selected reference curve. This characteristic is obvious in the case of the FWE criterion. The selection of a different off-axis angle to serve as reference curve can lead to very accurate, but also highly inaccurate, predictions as it was proved here and in [16, 20].

With the exception of the polynomial failure criteria, all the other examined fatigue theories are based on a number of assumptions that hinder if not prohibit their application for certain types of materials (e.g. HR cannot be applied when the failure modes of the material cannot be discriminated between fiber and matrix failure) or for certain loading patterns, such as the KW that was established only for fatigue loading patterns that do not include non-negative mean stresses.

6.6 Fatigue Design Considerations

The aim of the introduction and validation of multiaxial fatigue strength criteria, e.g., FTPF, FWE, is their potential practical application in the design of composite structures subjected to multiaxial, variable amplitude in general, fatigue stress states.

A composite structure may be composed of various laminate configurations. The examination of a number of fatigue theories in this chapter showed that two options are available if a multiaxial fatigue strength criterion is to be used in the design process; i) basic S–N curves of each lamina, which can be considered as the building layer of each laminate, must be known experimentally and laminate failure is theoretically predicted by means of a ply-by-ply analysis, as was assumed by the SB, KW and ST criteria, or ii) each laminate is considered as a different homogeneous anisotropic material whose basic strength properties must be measured experimentally, as assumed by the FTPF, FEW and HR fatigue theories. Obviously, in the first case only a limited number of tests have to be

Fig. 6.34 Definition of strength ratio for fatigue

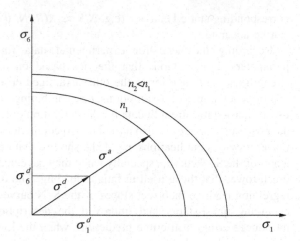

performed, but there is a lot of uncertainty regarding the remainder of the procedure, especially decisions concerning the degradation of ply properties after each ply failure.

If fatigue design is based on the second alternative, i.e., every multilayer configuration is considered to be a homogeneous anisotropic material with known basic S–N curves, and a polynomial-based fatigue failure criterion, e.g., FTPF, is used, the following comments apply.

6.6.1 Constant Amplitude Loading

In the special case where the developed stress fields are of constant amplitude, the dimensioning of the structure is a straightforward procedure. The strength ratio, SR, can be introduced for fatigue, having the same form as that for static loading [15]:

$$SR = \frac{\sigma_i^a}{\sigma_i^d} \tag{6.36}$$

where σ_i^a is the allowable or ultimate cyclic stress for the number of cycles for which cyclic design stress σ_i^d is imposed on the structure. If the applied stress components for n_1 cycles are only normal, σ_i^d, and shear, σ_6^d, components for example, the correlation of σ_i^a and σ_i^d is shown in Fig. 6.34. The value of the strength ratio is given by the positive root of the following quadratic equation:

$$SR^2 \left[F_{11}(n_1)\sigma_1^{d2} + F_{66}(n_1)\sigma_6^{d2} \right] + SRF_1(n_1)\sigma_1^d - 1 = 0 \tag{6.37}$$

In the above, it is assumed for simplicity that the stress ratio, $R = \sigma_{\min}/\sigma_{\max}$, of the design load is the same as that for which the basic S–N curves of the material are known. If material behavior must be theoretically simulated for different stress ratios than those available after material characterization, constant life diagram formulations like those presented in Chap. 4 must also be employed.

6.6.2 Variable Amplitude Loading

In general, multiaxial fatigue loads imposed on a structure are time series of variable amplitude and mean value; the same is true for the respective developed stresses. Assuming that multiaxial stress time series of normal and shear stresses are proportional to each other, a simple rainflow counting method such as that introduced in [32] can be used to convert these variable amplitude stress time series into blocks of data of constant amplitude and mean values. The use of constant life diagrams—like those described in Chap. 4—allows the substitution of these stress values by equivalent stress amplitudes corresponding to a mean value equal to that assumed for the experimental determination of the basic S–N curves.

By following the above procedure, the pairs of σ_{1i} and σ_{6i} amplitudes for a certain number of operating cycles, n_{oi}, are determined. Using FTPF or any other reliable fatigue failure criterion, the number of cycles to failure, N_{fi}, under each pair of σ_{1i} and σ_{6i} can be calculated by solving Eq. 6.16 for N for example, and a multiaxial Miner coefficient can be introduced:

$$D = \sum_{i=1}^{k} \frac{n_{oi}}{N_{fi}} \qquad (6.38)$$

where k is the total number of block loading patterns derived by the rainflow counting analysis.

6.7 Conclusions

The fatigue strength prediction of structures made of various composite material configurations operating under uniaxial or multiaxial loading conditions is the subject of this paper. The available failure criteria were examined and their predictive ability evaluated by the use of experimental data from the literature and the experimental program is presented in Chap. 2. These databases cover a wide range of GFRP and CFRP material configurations loaded under uniaxial and biaxial fatigue loads. The resulting complex stress fields are typical of those developed in the majority of engineering applications where composite materials are used.

The comparison between the evaluated criteria and the available experimental data showed that the polynomial fatigue failure criteria were very effective for estimation of the fatigue life of composite materials under complex fatigue stress states compared to those criteria based on one "master fatigue curve", which proved to be sensitive to the selection of this reference curve. However, considerably larger databases are necessary for the implementation of the polynomial criteria.

The accuracy of the examined multiaxial failure criteria in predicting the fatigue life of several composite material systems is generally acceptable. The application of the criteria proved to be easy and straightforward without the need for complicated numerical solutions. However, experience showed that the examined criteria can be accurate for one material and an examined loading case and highly inaccurate for another case. A commonly accepted complete fatigue theory does not yet exist, although some of the sub-problems have already been addressed.

The option to use laminate properties instead of lamina properties to predict laminate behavior enhances the applicability of the polynomial failure criteria (e.g. FTPF) for unidirectional and multidirectional lay-ups made up of any type of composite, e.g. unidirectional, woven or stitched layers. Especially under multiaxial loading, the polynomial failure theories (FTPF, SB) can produce acceptable fatigue failure loci for all the data considered, in contradiction with the FWE criterion, as used herein, which gives failure loci corroborated by the experimental multiaxial data only under certain conditions.

The efforts made with regard to fatigue failure prediction are aimed at more efficient design and consequently more accurate dimensioning of the parts of a structure. Using multiaxial strength criteria such as those described in this chapter, under constant amplitude loading, the lifetime is directly associated with the thickness of the laminate by using the fatigue strength ratio, SR. On the other hand, when the loading is irregular, a Miner coefficient for multiaxial stress states is introduced and the dimensioning is performed using an iterative procedure. The examined fatigue failure criteria can be used for investigation of the fatigue behavior of structural fiber-reinforced composite laminates under multiaxial stress states. Questions related to complex load proportionality, irregular stress spectra effects and potential validity of the constant life diagram formulations must be investigated before the proposed procedure can be used for realistic design cases.

References

1. Draft IEC 61400–1, Ed.2 (88/98/FDIS): 'Wind turbine generator systems–Part 1: Safety requirements', 1998
2. Germanischer Lloyd, 'Rules and regulations, IV–Non–marine technology', PART 1–WIND ENERGY, 1993
3. T.P. Philippidis, A.P. Vassilopoulos, Life prediction methodology for GFRP laminates under spectrum loading. Compos. Part A–Appl. Sci. **35**(6), 657–666 (2004)
4. J.A. Collins, *Failure of materials in mechanical design-analysis, prediction prevention.* (Wiley, New York, 1993)
5. R.F. Gibson, *Principles of Composite Material Mechanics* (McGraw-Hill Inc., New York, 1994)
6. O. Hoffman, The brittle strength of orthotropic materials. J. Compos. Mater. **1**(2), 200–206 (1967)
7. S.W. Tsai, E.M. Wu, A general theory of strength for anisotropic materials. J. Compos. Mater. **5**(1), 58–80 (1971)
8. Z. Hashin, A. Rotem, A fatigue failure criterion for fibre-reinforced materials. J. Compos. Mater. **7**, 448–464 (1973)

9. A. Rotem, Fatigue failure of multidirectional laminate. AIAA J. **17**(3), 271–277 (1979)
10. M.J. Owen, J.R. Griffiths, Evaluation of biaxial failure surfaces for a glass fabric reinforced polyester resin under static and fatigue loading. J. Mater. Sci. **13**(7), 1521–1537 (1978)
11. T. Fujii, F. Lin, Fatigue behavior of a plain-woven glass fabric laminate under tension/torsion biaxial loading. J. Compos. Mater. **29**(5), 573–590 (1995)
12. D.F. Sims,V.H. Brogdon, in *Fatigue Behavior of Composites under Different Loading Modes*, eds. by K.L. Reifsnider, K.N. Lauraitis. Fatigue of filamentary materials, (ASTM STP 636, 1977), pp. 185–205
13. M.-H.R. Jen, C.-H. Lee, Strength and life in thermoplastic composite laminates under static and fatigue loads. Part I: experimental. Int. J. Fatigue **20**(9), 605–615 (1998)
14. M.-H.R. Jen, C.-H. Lee, Strength and life in thermoplastic composite laminates under static and fatigue loads. Part II: Formulation. Int. J. Fatigue **20**(9), 617–629 (1998)
15. S.W. Tsai, H.T. Hahn, *Introduction to Composite Materials* (Technomic, Lancaster, 1980)
16. T.P. Philippidis, A.P. Vassilopoulos, Fatigue strength prediction under multiaxial stress. J. Compos. Mater. **33**(17), 1578–1599 (1999)
17. T.P. Philippidis, A.P. Vassilopoulos, Complex stress state effect on fatigue life of GRP laminates. Part II, Theoretical formulation. Int. J. Fatigue **24**(8), 825–830 (2002)
18. M. Kawai, A phenomenological model for off-axis fatigue behavior of unidirectional polymer matrix composites under different stress ratios. Compos. Part A-Appl. S **35**(7–8), 955–963 (2004)
19. Z. Fawaz, F. Ellyin, Fatigue failure model for fibre-reinforced materials under general loading conditions. J. Compos. Mater. **28**(15), 1432–1451 (1994)
20. M. Quaresimin, L. Susmel, R. Talerja, Fatigue behaviour and life assessment of composite laminates under multiaxial loadings. Int. J. Fatigue **32**(1), 2–16 (2009)
21. H. El Kadi, F. Ellyin, Effect of stress ratio on the fatigue failure of fiberglass reinforced epoxy laminae. Composites **25**(10), 917–924 (1994)
22. M.M. Shokrieh, F. Taheri-Behrooz, A unified fatigue life model for composite materials. Compos. Struct. **75**(1–4), 444–450 (2006)
23. R.S. Sandhu, R.L. Gallo, G.P. Sendeckyj, in *Initiation and Accumulation of Damage in Composite Laminates*, ed. by I.M. Daniel (ASTM STP 787, 1982), pp. 163–182
24. J. Awerbuch, H.T. Hahn, in *Fatigue of Fibrous Composite Materials*. ed. by K.N. Lauraitis. Off-axis fatigue of graphite/epoxy composites, (ASTM STP 723, 1981), pp. 243–273
25. S. Lee, M. Munro, Evaluation of in-plane shear test methods for advanced composite materials by the decision analysis technique. Composites **17**(1), 13–22 (1986)
26. S.W. Fowser, R.B. Pipes, D.W. Wilson, On the determination of laminate and lamina shear response by tension tests. Compos. Sci. Technol. **26**, 31–36 (1986)
27. A. Smits, D. Van Hemelrijck, T.P. Philippidis, A. Cardon, Design of a cruciform specimen for biaxial testing of fibre reinforced composite laminates. Compos. Sci. Technol. **66**(7–8), 964–975 (2006)
28. M.J. Hinton, A.S. Kaddour, P.D. Soden, *Failure Criteria in Fibre Reinforced Polymer Composites: The World-Wide Failure Exercise, a Composite Science and Technology Compendium* (Elsevier, Amsterdam, 2004)
29. T.P. Philippidis, P.S. Theocaris, Failure prediction of fibre reinforced laminates under hygrothermal and mechanical in-plane loads. Adv. Pol. Tech. **12**(3), 271–279 (1993)
30. A.P. Vassilopoulos, R. Sarfaraz, B.D. Manshadi, T. Keller, A computational tool for the life prediction of GFRP laminates under irregular complex stress states: Influence of the fatigue failure criterion. Comp. Mat. Sci. **49**(3), 483–491 (2010). 10.1016/j.commatsci.2010.05.039
31. R.P.L. Nijssen, O. Krause, T.P. Philippidis, Benchmark of lifetime prediction methodologies. Optimat Blades technical report, 2004, OB_TG1_R012 rev.001, http://www.wmc.eu/public_docs/10218_001.pdf
32. S.D. Downing, D.F. Socie, Simple rainflow algorithms. Int. J. Fatigue **4**(1), 31–40 (1982)

Chapter 7
Life Prediction Under Multiaxial Complex Stress States of Variable Amplitude

7.1 Introduction

Every engineering structure sustains numerous different loading patterns during its daily operation. Loading patterns of constant amplitude represent the minority, while the majority of the applied loads are of an irregular nature. A characteristic of the realistic loading is the stress multiaxiality, since, even when the external loads are of a uniaxial nature, the developed stress and strain fields are normally more complicated due to the material anisotropy.

The understanding and modeling of the material behavior under fatigue loading and the prediction of the fatigue life are difficult tasks, which become more complicated when the loading is of variable amplitude. There are two different classes of approaches that address this task [1]. The first is based on the selection of a damage summation rule for the fatigue life prediction of the examined material, without any observation of the actual damage mechanisms that develop in the material and cause the failure. The most commonly used is the linear Miner's damage accumulation rule because of its simplicity and the limited requirement for fatigue data. However, as it has been reported, e.g., [2, 3], fatigue life predictions for GFRP and CFRP composite materials are not accurate when this rule is used. Therefore, other, non-linear, damage accumulation rules were developed and applied for different material systems and loading conditions, e.g., [4, 5]. The second approach is based on the selection of a damage metric, such as the residual strength, e.g., [6–8], or the stiffness of the material, e.g., [9–11]. These methods can model the variance of the selected damage metric during fatigue life and eventually predict the lifetime based on predetermined failure criteria.

These theories were developed mainly based on constant amplitude and block loading fatigue data. However, over the last three decades, more fatigue data have been produced in laboratories concerning fiber-reinforced composite materials in order to examine their behavior under realistic loading situations, including variable amplitude fatigue loading and complex environmental conditions,

A. P. Vassilopoulos and T. Keller, *Fatigue of Fiber-reinforced Composites*,
Engineering Materials and Processes, DOI: 10.1007/978-1-84996-181-3_7,
© Springer-Verlag London Limited 2011

e.g., [12–18]. In parallel to the experimental work that is performed, theoretical models are developed for the simulation of the fatigue behavior of the examined materials under different thermomechanical loading conditions and the prediction of material fatigue life under complex stress states that may arise during the operation of a structure in the open air.

A fatigue life prediction methodology is usually based on the development of empirical relationships between the applied loads and the fatigue lifetime of the examined materials. The implementation of a numerical procedure for fatigue analysis consists of a number of distinct calculation modules, related to life prediction. Some of these are purely conjectural or of a semi-empirical nature, e.g., the failure criteria, while others rely heavily on experimental data, e.g., S–N curves and constant life diagrams (CLDs). In cases of composite laminates under uniaxial loading, leading to uniform axial stress fields, the situation can be substantially simplified since almost all relevant procedures could be implemented by experiment. On the other hand, for complex stress states, the laminated material is considered as being a homogeneous orthotropic medium and its experimental characterization, i.e., static and fatigue strength, is performed for both material principal directions and in-plane shear [19]. This *laminate approach* is a straightforward one for predicting fatigue strength under plane stress conditions, avoiding the consideration of damage modeling and interaction effects between the plies and stress redistribution, and can be reliably used when limited stacking sequence variations are present in a structural element. The approach was implemented by Philippidis and Vassilopoulos [17, 20–22] for a glass/polyester multidirectional laminate of [0/±45] stacking sequence and was shown to yield satisfactory predictions for fatigue strength under complex stress conditions for both constant amplitude and variable amplitude loading. A straightforward algorithm must be followed comprising at steps dealing with the analysis of the load to determine the developed stress fields, the interpretation of the fatigue data of the examined laminates to derive the S–N curves and the corresponding constant life diagrams, the fatigue failure criteria of the calculation of the design allowables and finally the damage accumulation based on the selected damage rule.

In large composite structures, consisting of numerous different materials and laminate configurations, a *lamina-to-laminate* approach seems more appropriate, although requiring the development of additional calculation modules that are able to take into account the implications in local stress fields, stress redistribution in neighboring plies and finally, how damage propagates as a function of loading cycles, e.g., [19, 23, 24]. In cases like this, the material properties of the basic building ply need to be experimentally derived. The properties of any new laminate configuration are then estimated based on existing theoretical procedures. The failure analysis is based on a progressive damage modeling, considering failed layers and stress redistributions in the laminate according to the applied load history.

It is obvious that whatever method will be followed for the life estimation, implementation of the necessary steps into a computer program is indispensable. Software products have been developed to assist the research community and simplify design processes. The methods implemented in the software packages were based either on the micromechanics modeling of fatigue life, e.g. Helius:Fatigue [25] or the phenomenological laminate approach described above, e.g. *CCfatigue* [26].

In this chapter, a classic fatigue life prediction methodology (based on the laminate approach) is presented and applied to the constant and variable amplitude fatigue experimental data from Chap. 2 in order to examine the accuracy of the predictions. The effect of the various parameters of the proposed methodology on the final life prediction result is investigated. The application of the methodology is assisted by the recently developed computational tool *CCfatigue* [26]. *CCfatigue* is a modular software framework that allows rapid interchange between modules in a complete fatigue life prediction methodology and the evaluation of their effect on the life prediction results. A software library has already been developed, containing a significant number of solutions for each step of the fatigue life prediction methodology.

7.2 Classic Fatigue Life Prediction Methodology

The estimation of fatigue life based on classic fatigue life prediction methodology comprises at a number of sequentially executed modules. Therefore, a classic life prediction methodology can be considered as an articulated method consisting of four to five basic steps:

- Cycle counting, to convert variable amplitude (VA) time series into blocks of certain numbers of cycles corresponding to constant amplitude and mean values.
- Interpretation of fatigue data to determine and apply the appropriate S–N formulation for the examined material.
- Selection of the appropriate formulation to take the mean stress effect on the fatigue life of the examined material into account.
- Use of the appropriate fatigue failure criterion to calculate the allowable number of cycles for each loading block that results after cycle counting.
- Calculate the sum of the partial damage caused to the material by each of the applied loading blocks estimated using the cycle-counting method.

Modeling constant amplitude fatigue behavior involves the determination of the S–N curves (plot of cyclic stress vs. life), typically by grouping data at a single R-value ($\sigma_{min}/\sigma_{max}$) (incidentally, depending on the R-value, an S–N curve can be constructed from data obtained at varying mean and amplitude). Interpretation of the fatigue behavior for the assessment of the mean stress effect results in the construction of the constant life diagram (CLD). These two processes can be treated as separate steps, but are related in the sense that the CLD is constructed from the available S–N curves, and new S–N curves could be extracted from this CLD.

The flowchart presented in Fig. 7.1 shows the procedure followed in order to calculate the damage index for a given variable amplitude fatigue problem. Information regarding the 2nd, 3rd and 4th steps that concern the interpretation of

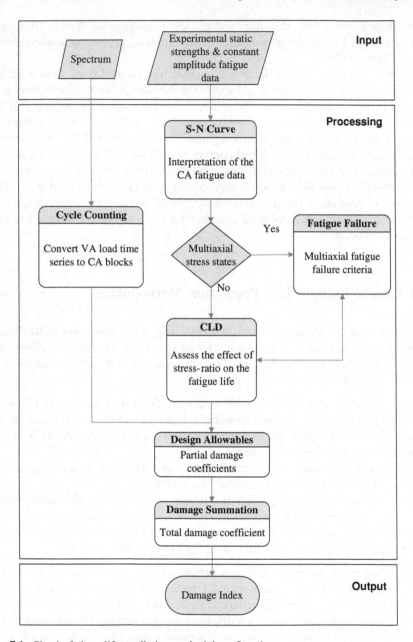

Fig. 7.1 Classic fatigue life prediction methodology flowchart

the fatigue data (Chaps. 3 and 4), constant life diagrams (Chap. 4) and fatigue failure criteria (Chap. 5) can be found earlier in this volume. The cycle-counting methods and damage accumulation rules will be briefly described here.

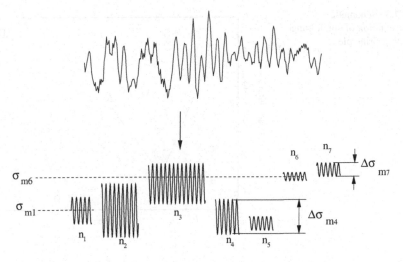

Fig. 7.2 Schematic representation of application of a cycle counting method

7.2.1 Cycle Counting

Cycle counting is used to summarize (often lengthy) irregular load vs. time histories by providing the number of occurrences of cycles of various sizes, see Fig. 7.2. The definition of a cycle varies with the method of cycle counting. A significant number of cycle-counting techniques have been proposed over the last 30 years, e.g., [27]. Typical events observed in a load-time history are the occurrence of load peaks or valleys at specific levels, the exceeding or "crossing" of specific levels and the occurrence of load changes or "ranges" of a specific size. Accordingly, cycle-counting methods can be classified as follows:

- Level-crossing counting methods.
- Peak counting methods.
- Simple range-counting methods and
- Rainflow counting methods.

Level-crossing, peak counting and simple range counting are designated one-parameter methods since they result in the counting of solely one load parameter. On the other hand, range-mean and rainflow-related methods are designated two-parameter methods and are more appropriate for the fatigue analysis of composite materials since they count how many times a cycle of specific range and mean value occur in a spectrum, thus taking into account the effect of mean stress on the fatigue life of the examined material.

The application of a cycle-counting method and storage of its results require a preliminary treatment of the loading, which consists of sampling, extraction of maximum and minimum loads and finally quantification of the values into classes.

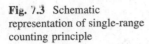

Fig. 7.3 Schematic representation of single-range counting principle

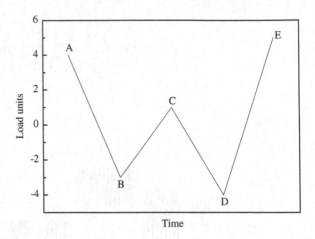

Sequences are generally produced by the monitoring of a continuous variable over time. Signal processing is then used to convert analog signals into digital ones, which consist of discrete digital values. Sampling is necessary because the digital signal has to correctly represent the evolution of the variable. Sequences of force, stress, strain or other loading parameters can be stored. These sequences are known as load-time histories.

Cycle-counting methods consider only the successive extremes of a load-time history and the sampled sequence is therefore reduced to a peak/valley sequence before cycle counting. Depending on which method is used, peaks or valleys or both may be required. In the classification of peaks and valleys a specific "range filter" is usually applied to remove low-range cycles from the sequence. The reversal points remaining after the search for peaks and valleys are discarded. Cycle-counting algorithms are then applied to the reduced load-time history, which count ranges as cycles using counting principles.

Two main principles can be applied for the cycle counting: the single-range or range-mean counting and the range-pair counting principle, on which rainflow counting is based as well. According to the single-range counting, Fig. 7.3, any transition from a peak, e.g. A, to a valley e.g. B, in a time series or from a valley B to a peak C, forms one half cycle with range = |A–B| and mean = (A + B)/2, or range = |B–C| and mean = (B + C)/2. Counting in single ranges results in as many half cycles as the number of load transitions in the loading sequence.

The rainflow counting and the earlier range-pair methods are used to extract hysteresis loops from a load history. The algorithm of counting cycles, based on the comparison of four successive data points of the loading history, is schematically shown in Fig. 7.4. The concept of stress–strain hysteresis loops resulting from a certain load history is visualized in Fig. 7.5.

A comprehensive description of the available cycle-counting methods for the analysis of spectra applied to composite materials is given in [28]. As mentioned in [28] the history of cycle-counting methods goes back to the 1950s and 1960s when

Fig. 7.4 Schematic representation of range-pair counting principle

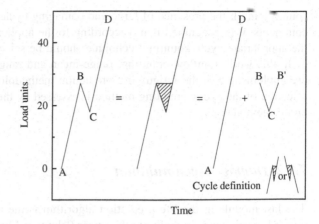

Fig. 7.5 Stress–strain hysteresis loops resulting from rainflow counting of an irregular spectrum

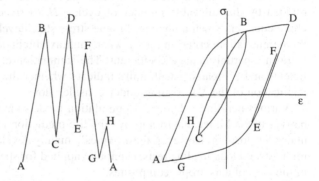

only simple range-counting or range-mean counting methods were used. The drawback of these methods is their inability to take into account the stress–strain history to which the material is subjected, and consequently their tendency to miss the largest overall load cycle in a sequence. The rainflow counting and related methods were introduced to address this problem [29–31]. Based on these algorithms, stress–strain hysteresis loops are counted rather than stress range and mean values. Endo et al. [29, 30] in Japan developed an algorithm that became known as the rainflow or "pagoda-roof" counting method. At the same time, in Europe, a two-dimensional method was developed by de Jonge [31] designated the "range-pair-range" or just "range-pair" method. Although based on different backgrounds, both cycle-counting algorithms yielded the same results when counting an arbitrary loading sequence.

Rainflow-counting, range-pair and range-mean methods seem the most appropriate for the analysis of composite material fatigue data, giving similar cycle-counting results for most practical applications [32]. However they present a number of deficiencies: rainflow-counting cannot be used for cycle-by-cycle analysis and it is therefore difficult to apply this method in combination with a residual strength fatigue theory. On the other hand, range-pair and range-mean counting mask the presence of large and damaging cycles. Based on the previous comments, it is concluded that—according to the application and material used– the appropriate cycle-counting technique should be selected very carefully.

In this work, rainflow-counting, range-mean and range-pair counting methods are implemented in the fatigue life prediction methodology. The influence of the selection of the cyclic counting method is assessed by the predictive ability of the entire methodology.

7.2.2 Damage Accumulation

The last module in the life prediction algorithm is the accumulation of damage, which is carried out according to the linear Palmgren-Miner or the Miner rule. The number of operating cycles, n, of each bin, derived from rainflow counting, is divided by the allowable number of cycles, N_f (derived directly using the S–N curve equation, when a uniaxial fatigue stress state develops or by the solution of the fatigue failure criterion for N_f, when multiaxial fatigue stress fields are present) to form a partial damage coefficient. The summation of all partial damage coefficients and comparison with unity indicate whether the material will survive the application of the VA loading under consideration.

Various non-linear damage accumulation models have been proposed as alternatives to the Miner rule to improve the life prediction for anisotropic composites under VA loading. However, these "rules", mainly developed for two-stage and/or multi-stage block loading tests, cannot be applied for spectrum loading, especially in the case of anisotropic composites.

The difference between the Miner and non-linear damage summation rules is schematically presented in Fig. 7.6. For the linear damage rule, the damage caused to the material is the same for the same number of applied cycles and there is no relation to the history, i.e., the amount of existing damage to the material. On the other hand, when a non-linear rule is selected (simple cases are used for the example in Fig. 7.6) the damage accumulation is not linear, depending on rule type.

Non-linear "rules" yield more accurate life predictions than the Miner rule, simply because they are fitted to VA experimental data. They cannot be used for design purposes where numerous different load cases, composed of different spectra, must be examined. In this respect, they are not really damage accumulation rules.

An extensive review of existing cumulative damage models, not only for metals but also for composite materials, can be found in [33]. One of the simpler non-linear rules was introduced by Marco and Starkey [2] and assumes that damage accumulation is a function of the cyclic stress level:

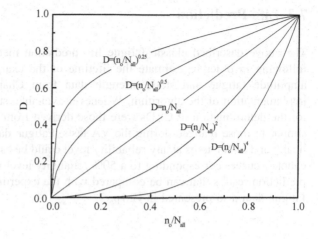

Fig. 7.6 Schematic representation of linear and non-linear damage accumulation rules

$$D = \sum_{i=1}^{k} \left[\left(\frac{n_{oi}}{N_{fi}} \right)^{a_i} \right] \tag{7.1}$$

where the parameter a_i is a function of stress level and must be estimated based on available experimental data.

Owen and Howe [5] developed a stress-independent non-linear cumulative damage model for chopped strand mat GRP, given by:

$$D = \sum_{i=1}^{k} \left[A \left(\frac{n_{oi}}{N_{fi}} \right) + B \left(\frac{n_{oi}}{N_{fi}} \right)^{2} \right] \tag{7.2}$$

A modification of this model with the quadratic exponent replaced by another variable parameter was presented in [3]:

$$D = \sum_{i=1}^{k} \left[A \left(\frac{n_{oi}}{N_{fi}} \right) + B \left(\frac{n_{oi}}{N_{fi}} \right)^{c} \right] \tag{7.3}$$

In the above equations, n_{oi} and N_{fi} are the numbers of operating cycles and cycles to failure, respectively of the ith bin. In [2], the three parameters, A, B and c, were calculated using an iterative procedure to fit Eq. 7.3 to the experimental data. The same model, Eq. 7.3, was also used for the life prediction of CFRP angle-ply and pseudo-isotropic laminates tested under the FALSTAFF (Fighter Aircraft Loading STAndard For Fatigue) spectrum [34] and to predict the variable amplitude fatigue behavior of glass-fiber reinforced polyester laminates used for the construction of wind turbine rotor blades [17].

7.3 Life Prediction

The above described classic fatigue life prediction methodology is used in the following in order to estimate the lifetime of the examined material. Constant amplitude fatigue and static strength data from Chap. 2 were used for the implementation of the algorithm. Whenever available, static strength values used for the determination of CLDs were those derived from static tests at strain rates similar to those realized during the VA tests. Fatigue data were processed statistically and S–N curves at any reliability level could be estimated. However, in this chapter, curves corresponding to a 50% reliability level were used to produce life prediction results that can be compared with the experimental variable amplitude fatigue data also presented in Chap. 2.

7.3.1 Configuration of the Classic Life Prediction Methodology: Presentation of CCfatigue Software

Several modules were developed for the solution of each of the successive steps of the previously described classic fatigue life prediction methodology. Four cycle-counting techniques were implemented in the library of the *CCfatigue* software, four S–N curve formulations, five different methods for the construction of constant life diagrams, six multiaxial fatigue failure criteria for the calculation of the allowable number of cycles under the developed stress states and two methods for the damage accumulation: the linear Miner's rule and the non-linear equation described in the previous sections, Eq. 7.3, for the summation of the accumulated damage and assessment of the lifetime of the examined material. The *CCfatigue* framework is thus capable of predicting the fatigue life of a number of different material systems under different loading conditions. The flowchart of the methodology is presented in Fig. 7.1, which schematically shows the process used for predicting the Miner's damage index for a variable amplitude fatigue problem. The five distinctive steps of the methodology are presented in this section showing the data processing part of the software framework. The current version of *CCfatigue* offers 1,344 solver combinations, showing its capacity to address a wide range of materials and loading combinations.

The software framework and the existing modules are briefly introduced in this paragraph. The home page is shown in Fig. 7.7 and the five successive steps can be seen on the left hand side of the above screen shot. The user can run each step independently of the others, select the desired solver from those available in the software library and obtain the solution. The data structure of the output file for each step has been designed in such a way that it can be used as the input file for the following step. The implementation of the software for the life prediction of the selected materials and evaluation of the selected modules for each step are presented in the following sections.

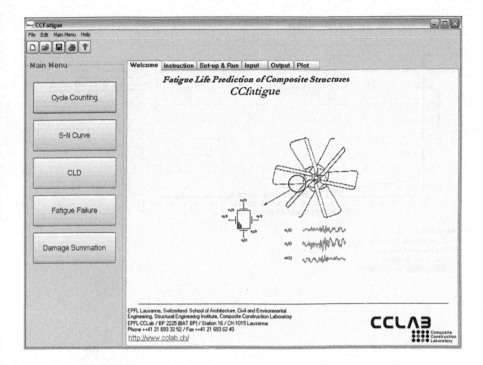

Fig. 7.7 *CCfatigue* software home page

A demonstration of the application of the software to solve each step of the methodology using different solvers and a brief discussion of their effect on the modeling of the exhibited fatigue behavior of the examined material under the given variable amplitude follows.

Cycle counting of both irregular spectra, presented in detail in Chap. 2, can be performed following the algorithms of the four methods implemented in the *CCfatigue* software. The cumulative spectra of the MWX and EPET573 spectra presented in Figs. 7.8 and 7.9, respectively show that there is no significant discrepancy if different methods are employed for counting the loading cycles. In general, all methods count the same or at least similar numbers of cycles, except range-mean which counts more cycles than the other three. In addition, the cumulative spectrum resulting from the application of the range-mean counting method is slightly different from the others, presenting more low-range cycles than the rainflow, simplified rainflow and range-pair counting algorithms.

The application of different S–N curve methods is demonstrated in Fig. 7.10 on fatigue data recorded after the application of compression fatigue loading ($R = 10$) to on-axis specimens. As shown, all S–N curves model the exhibited fatigue behavior similarly in the region between 10^3 and 10^6 cycles, but they are significantly different in the low- and high-cycle fatigue regimes. Depending on the cycle-counting results, each method can produce a conservative on

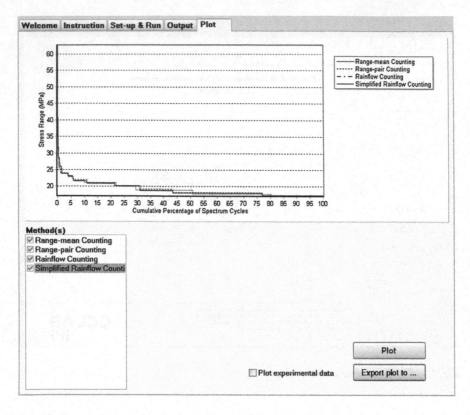

Fig. 7.8 Comparison of cumulative spectra resulting from different cycle-counting routines; MWX time series

non-conservative lifetime estimation. If the cumulative spectrum contains a significant percentage of high-range cycles for example, corresponding to low-cycle fatigue, use of Lin-Log or Log-Log relationships will yield conservative lifetime estimations, while use of the Sendeckyj method will yield non-conservative lifetime estimations.

The selection of the CLD also affects the final fatigue life prediction in a similar manner. As presented in Fig. 7.11, different CLD methods produce different diagrams and eventually estimate different S–N curves for desired stress ratios. Therefore, depending on the accuracy with which the CLD formulation estimates an S–N curve, fatigue life prediction can be accurate, conservative or non-conservative. The application of three methods—piecewise linear, Boerstra and piecewise non-linear (see Chap. 4 for explanations)—to the examined composite material system of Chap. 2 is demonstrated in Fig. 7.11. As shown, piecewise linear and non-linear diagrams are more conservative than Boerstra and therefore the use of one of these solutions would eventually yield more conservative fatigue life prediction results.

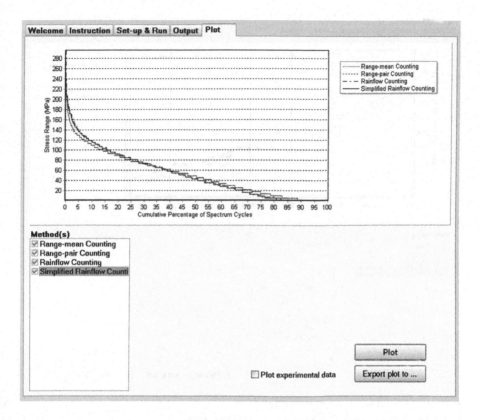

Fig. 7.9 Comparison of cumulative spectra resulting from different cycle-counting routines; EPET573 spectrum

In cases where a uniaxial stress state develops in the examined material, the next step, related to fatigue failure criterion selection can be omitted as selection of the appropriate S–N curve and CLD formulations suffices to take the effect of any possible variable amplitude loading pattern into account.

However, when the loading or resulting stress field is complex, constant life diagrams must be available for all strength parameters of the material in its symmetry directions, and the aforementioned interpolation procedure must be performed separately for all the plane stress tensor components. Their interaction in determining failure is then taken into account by considering multiaxial fatigue strength criteria. Six models were implemented in *CCfatigue* to cover as many concepts that deal with this problem as possible. Three of the selected models, Hashin-Rotem (HR), Sims-Brogdon (SB) and Failure Tensor Polynomial in Fatigue (FTPF)—see Chap. 6 for details—can be characterized as the macroscopic fatigue strength criteria, which are usually generalizations of known static failure theories for taking factors relevant to the fatigue life of the structure into account, such as number of cycles and loading frequency. Two more criteria, one proposed by Kawai (KW) and the second by Fawaz and Ellyin (FWE), seem very promising

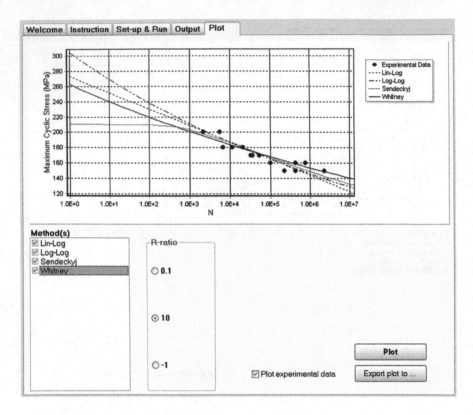

Fig. 7.10 Different types of S–N curves produced by *CCfatigue*

since they include the stress ratio in their formulation and can therefore be used without the need for any CLDs and are consequently based on limited databases. The sixth model, introduced by Shokrieh and Taheri (ST), is based on a strain energy concept and its use is also limited to the modeling of the off-axis fatigue behavior of unidirectional materials.

Although other models can be found in the literature, those selected here have been proved more accurate than others—see Chap. 6—and can be easily implemented in *CCfatigue* and fatigue databases that exist in the literature for the evaluation of their accuracy. The *CCfatigue* interface for selection of the appropriate fatigue failure theory and input of the requested parameters is presented in Fig. 7.12 (the HR criterion is used for the demonstration). The static-tensile and compressive strengths of the material are required as input together with three S–N curves (normally those corresponding to the fatigue functions of the material along the longitudinal and transverse directions and under shear). Use of the selected criterion allows the prediction of the S–N curve for the examined material system at any off-axis angle.

Damage summation is performed during the last step of the *CCfatigue* software framework—a typical "Set-up & Run" tab is presented in Fig. 7.13. The

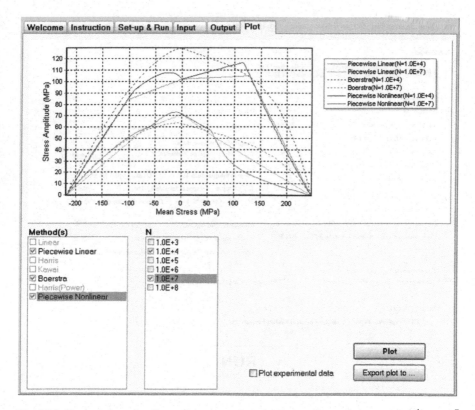

Fig. 7.11 Typical piecewise linear, Boerstra and piecewise non-linear diagrams for 10^4 and 10^7 cycles derived using *CCfatigue*

combination of the solvers for the solution of each step of the articulated methodology can be selected in the interface. A fatigue damage index is the output of this module, corresponding to the damage accumulated in the material after application of the selected fatigue spectrum. In the present version of the software, the linear Miner rule and the non-linear equation described earlier Eq. 7.3 were implemented.

7.3.2 Application of the Classic Fatigue Life Prediction Methodology

The classic fatigue life prediction methodology was applied to the fatigue data of the examined material in order to evaluate its predictive ability. The theoretical predictions were compared to variable amplitude experimental data under the modified version of the standardized WISPERX spectrum, designated MWX in Chap. 2, and an alternative variable amplitude spectrum, designated EPET573.

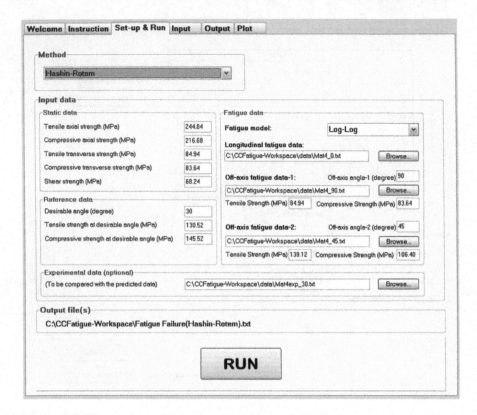

Fig. 7.12 *CCfatigue* interface for setting-up of fatigue failure solver

The following configuration of the *CCfatigue* software framework was selected with the chosen solvers being the most common for each step:

- Rainflow cycle-counting method,
- Log–Log S–N curve formulation,
- Piecewise linear constant life diagram,
- Miner's damage rule.

In cases where multiaxial stress states developed in the material, the Failure Tensor Polynomial in Fatigue (FTPF), described in Chap. 6, was employed to calculate the allowable number of cycles for the developed stress state.

A comparison of theoretical predictions with experimental data is presented in Figs. 7.14, 7.15, 7.16 and 7.17. Experimental data points from variable amplitude tests with MWX are compared with theoretical predictions, shown as solid lines, in Fig. 7.14 for on-axis and in Fig. 7.15 for off-axis tests. In general, the theoretical numbers of spectrum passes are found to be in satisfactory agreement with the experimental values. Nevertheless, this is not the case for specimens tested under the alternative spectrum EPET573, Figs. 7.16 and 7.17, for which conservative predictions resulted, especially those regarding the on-axis data. Therefore it can

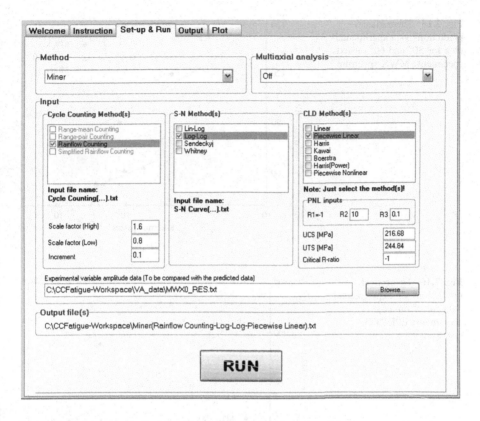

Fig. 7.13 Fatigue life prediction estimation based on selected modules

Fig. 7.14 Theoretical predictions vs. experimental data for 0° on-axis specimens; MWX spectrum

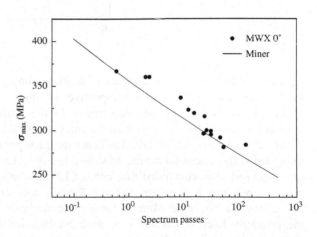

be concluded that the predicting efficiency of the methodology, in the present framework, is not affected by fiber lay-ups, but by the spectrum itself. For the more "irregular" EPET573 spectrum, results are less accurate, while for the MWX

Fig. 7.15 Theoretical predictions vs. experimental data for 30° and 60° off-axis specimens; MWX spectrum

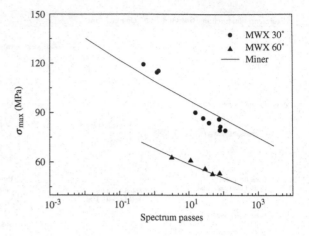

Fig. 7.16 Theoretical predictions vs. experimental data for 0° on-axis specimens; EPET573 spectrum

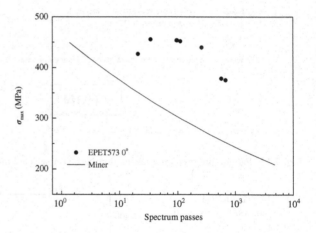

spectrum, which is almost a sequence of block loading patterns, relatively accurate predictions can be achieved irrespective of fiber orientation. Discrepancies observed between theory and experiment in the latter case could possibly be caused by any one of the algorithm modules described.

Since the fatigue failure criterion is not used for predicting the response of on-axis loaded specimens for which, as shown in Fig. 7.16, the greatest discrepancies are observed, it is concluded that either CLD resolution, implemented by inter-polating between S–N curves at three R-ratios and static strengths, or the S–N curve formulation, or the Miner's damage accumulation rule could be the cause of the problem. Each of the above or their combination can drastically affect pre-diction accuracy. The S–N curves, for example, were used successfully for the life prediction of specimens subjected to the MWX spectrum, see Fig. 7.14, but their validity is questioned when the EPET573 spectrum is used, see Fig. 7.16. This could be explained by the fact that MWX is a sequence of blocks of constant

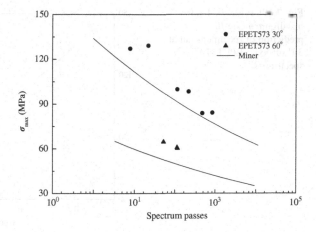

Fig. 7.17 Theoretical predictions vs. experimental data for 30° and 60° off-axis specimens; EPET573 spectrum

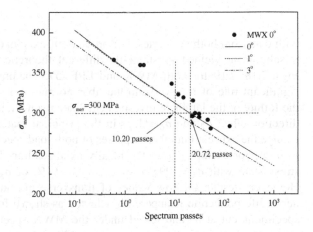

Fig. 7.18 Effect of load misalignment on life prediction of specimens cut at 0° on-axis; MWX spectrum

high-range cycles and that the present S–N formulation is found to be adequate. On the other hand, the EPET573 spectrum consists of many low-range cycles, see Fig. 7.20 in Chap. 2, and for these stress ranges the S–N curves must be extrapolated far beyond 10^7 cycles to determine the allowable number of cycles. However, the derived S–N curves are extracted from test data between approximately 10^3 and 3×10^6 loading cycles and are obviously not representative for the region beyond 10^7 cycles.

The damage accumulation rule and associated rainflow counting, which implies the use of Miner's rule, could also affect life prediction; by comparing the two variable amplitude time series (Figs. 2.17 and 2.18 in Chap. 2), and their cumulative spectra (Figs. 7.8 and 7.9) clearly show that the application of EPET573, consisting of alternating high- and low-range cycles, would result in pronounced load sequence effects.

Regardless of the effect of various parameters on the accuracy of the theoretical predictions, a strong effect of the plane stress state was observed in comparisons

Fig. 7.19 Effect of load misalignment on life prediction of specimens cut at 30° off-axis; EPET573 spectrum

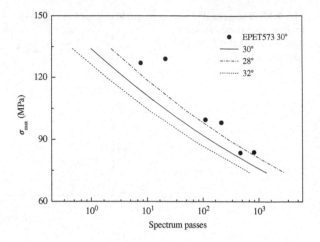

with data from both test series. For on- and off-axis loading, even a few degrees of misalignment yielded substantially different theoretical predictions as shown in Figs. 7.18, 7.19 for both MWX and EPET573 loading spectra. This proves the significant role of transverse-to-the-fiber normal and shear stress components in the failure of the laminate investigated, since even slight deviations of the loading direction affect σ_2 and σ_6 values in the principal material system and drastically change the expected number of passes of both load spectra. A misalignment of two degrees, for example, from the ideally axial loading direction, produces a plane stress state with $\sigma_1 = 99.9\%$ of σ_x, $\sigma_2 = 0.1\%$ of σ_x and $\sigma_6 = 0.035\%$ of σ_x. However, despite the low values of transverse normal and shear stress components, life prediction is apparently affected as shown for example in Fig. 7.18 for specimens cut at 0° and tested under the MWX spectrum and also in Fig. 7.19 where predictions for specimens cut at 30° and tested under EPET573 loading spectrum are shown.

Care should be also taken regarding the selection of the data used as baseline. Different damage coefficients would be calculated if Log-Log or Lin-Log curves are used for S–N data representation as mentioned in [16]. Moreover, even UTS or UCS values, used for determination of the constant life diagrams, could drastically affect the expected lifetime. As was mentioned earlier, for the construction of the CLDs, ultimate stress values derived at strain rates similar to those realized in the VA series were used, wherever available. These values were significantly greater than corresponding values determined in static tests, performed according to relevant standards, and indeed, as shown in Figs. 7.20, 7.21, they affect the expected life predictions for both loading spectra. As can be seen from Fig. 7.20, the use of the UTS derived at a high strain rate leads to the accurate prediction of the fatigue life of specimens cut at 0° and tested under the MWX spectrum, while use of a lower value of the UTS derived at a standard test speed yields conservative predictions. The difference in predictions also exists when the fatigue behavior of

Fig. 7.20 Effect of ultimate tensile stress on life prediction results. MWX spectrum applied on-axis (HSR: high strain rate, SSR: standard strain rate)

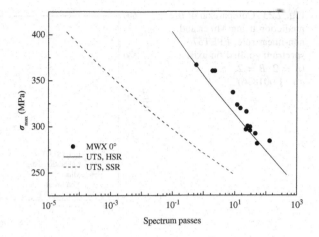

Fig. 7.21 Effect of ultimate tensile stress value on life prediction results. EPET573 spectrum applied on-axis (HSR: high strain rate, SSR: standard strain rate)

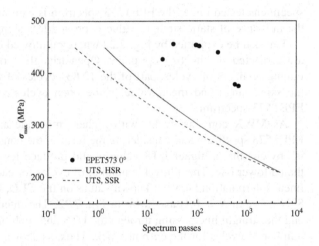

Fig. 7.22 Calculation of allowable number of cycles for selected bins of both spectra using different UTS values

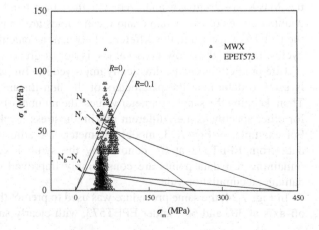

Fig. 7.23 Comparison of life prediction using Miner and non-linear rule. EPET573 spectrum applied on-axis. ($A = 2$, $B = 2$, $c = 1.001826$)

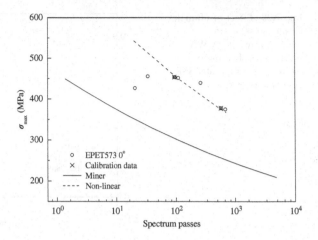

specimens tested under the EPET573 spectrum is examined, although in this case the influence of static strength value is proved less significant, see Fig. 7.21.

This can be explained by Fig. 7.22 where rainflowed data from both time series are projected on the (σ_m–σ_a)-plane. Open triangles correspond to blocks, containing numbers of cycles, part of the 12,831 cycles of the MWX spectrum, with the same range and mean value, while open cycles refer to CA blocks of the EPET573 spectrum.

As MWX contains cycles with higher amplitude and mean values than the EPET573 spectrum, and considering the form of the constant life diagrams used, it is obvious that a higher UTS value would produce less conservative predictions than a lower one. The allowable number of cycles for each block is determined by linear interpolation between known values on the CLD, for example, UTS and the S–N curve at $R = 0.1$, as shown in Fig. 7.22. The intersection of the S–N curve and the straight line passing through the UTS and each block defines the allowable number of cycles for the current block. Thus, as shown in Fig. 7.22, significantly different numbers of allowable cycles are calculated when rainflowed data from the MWX spectrum are analyzed. On the other hand, it is obvious that when counted cycles of lower range and mean values are examined, typical of those of the EPET573 spectrum, the difference between the calculated allowable number of cycles, using high or low UTS values, is not as great.

Life predictions can be drastically improved if a limited number of VA test data is used to determine the parameters of the non-linear fitting equation (Eq. 7.3). Then, keeping the same parameter values, the number of spectrum passes to failure for other specimens, i.e., different maximum stress levels, is accurately predicted. For example, in Fig. 7.23, model parameters are adjusted by using experimental data from EPET573 at $0°$, denoted by the symbol ×, and predictions for the remaining test data points are considerably improved compared with the linear damage accumulation rule.

In Fig. 7.24, the same procedure was used to predict the life of specimens tested off-axis at $30°$ and $60°$, under EPET573, with clearly satisfactory results.

Fig. 7.24 Comparison of life prediction using Miner and non-linear rule. EPET573 spectrum applied on 30° (A = 1.5, B = −1.5, c = 1.0529) and 60° (A = 1.5, B = −1.5, c = 1.0127) off-axis specimens

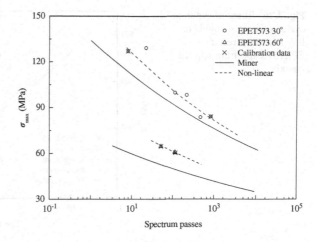

Fig. 7.25 Effect of different cycle-counting methods on life prediction results. MWX spectrum applied on-axis

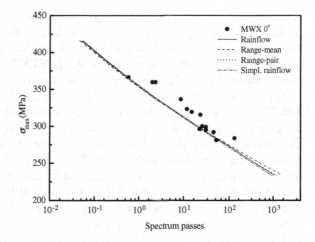

7.3.3 Alternative Configurations

Although the presented results are based on a simple sequence of solvers selected for the solution of each step of the classic fatigue life prediction methodology, alternative configurations can also be selected for determination of the damage coefficient in order to investigate the effect of each solver on life prediction results.

7.3.3.1 Effect of Cycle-Counting Method

As shown in Fig. 7.25 for the MWX spectrum, predictions using the classic fatigue life methodology are corroborated by the experimental data, irrespective of the cycle-counting algorithm used. However, this is not the case for the EPET573

Fig. 7.26 Effect of different
cycle-counting methods on
life prediction results.
EPET573 spectrum applied
on-axis

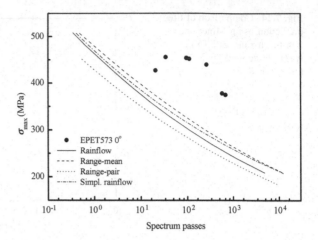

spectrum, as shown in Fig. 7.26. As previously mentioned, see Fig. 7.16, the life prediction for this spectrum using the standard configuration was not accurate. Nevertheless, as presented in Fig. 7.26 use of the range-mean cycle-counting technique improves the predictions for the EPET573 spectrum compared to the results obtained from the application of the other (rainflow-related) techniques.

One reason for this could be the extreme and frequent load transitions in the EPET573 time series and the characteristic of rainflow counting-related methods to "produce" cycles that do not really exist, as they conjugate unrelated parts from throughout the time series. Thus, when the time series consists of extreme and frequent load transitions, such as EPET573, the phenomenon is intensified and differences in the numbers of cycles counted by rainflow-related and single range-mean methods are more evident. For the smoother MWX time series however, this is not the case, as a considerable number of cycles counted using rainflow-related methods are formed from adjacent load reversals.

It is demonstrated that all examined cycle-counting methods can be used as part of the entire life prediction methodology and provide more or less accurate results depending on the examined material and applied loading spectrum. It should nevertheless be noted that these comments must be validated by additional comparisons between predictions and other experimental data, considering different loading spectra and/or different material systems.

7.3.3.2 Effect of S–N Curve Formulation

The selection of the S–N curve formulation was also found to affect the life prediction results for the examined material and the applied variable amplitude spectra. As shown in Fig. 7.27, the two formulations based on the Log-Log equation (the linear log S vs. log N regression and the method proposed by Whitney) provide similar and quite accurate results. The semi-logarithmic (Lin-Log) equation

Fig. 7.27 Effect of different S–N formulations on life prediction results. MWX spectrum applied on-axis

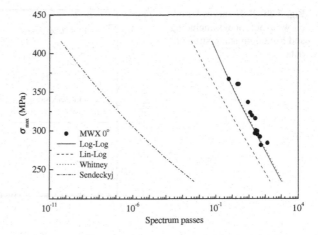

Fig. 7.28 Modeling of low-cycle fatigue regime by different S–N curve formulations

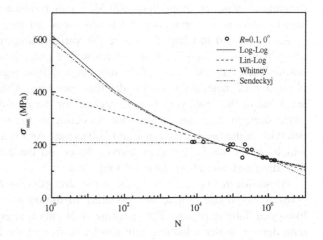

is less accurate as it becomes conservative, while the prediction from the Sendeckyj wear-out model is far less accurate when estimating failure at very early stages of loading.

In Fig. 7.28 the S–N curves derived by the different methods are presented for the R-ratio of 0.1, the closest available to the average stress ratio observed in the MWX variable amplitude spectrum. As shown in this figure, the S–N curves based on the Log-Log formulation overestimate the life in the low-cycle fatigue regime, estimating the tensile strength of the material at a value of around 600 MPa, roughly 1.5 times higher than the experimentally derived static strength obtained at a high loading rate (see Chap. 2). The S–N curve derived based on the semi-logarithmic equation models well the behavior in the low-cycle regime, but underestimates the behavior for high-cycle fatigue. The curve derived by the wear-out model without using the static data however is very conservative in the low-cycle fatigue regime. According to this S–N curve, even one cycle with a

Fig. 7.29 S–N curves based on wear-out model, including and excluding static strength data

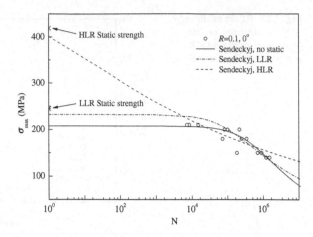

maximum stress of more than 200 MPa can produce failure of the examined on-axis specimens, something that is obviously not true.

As mentioned in Chap. 2—where the variable amplitude spectra were analytically presented—MWX has no cycles with a range lower than 30% of its maximum. Therefore, modeling of the low-cycle fatigue regime is critical for the life prediction of materials loaded under this spectrum. Although no fatigue data are available in the low-cycle fatigue regime for the examined material, use of the static strength data can improve the modeling of the constant amplitude fatigue behavior in the range between 1 and 100 cycles. An example is given in Fig. 7.29, which shows the S–N curves derived based on the Sendeckyj wear-out model including and excluding static strength data.

As shown in Fig. 7.29 using the static strength data considerably improve the constant amplitude fatigue life modeling of the on-axis loaded specimens in the low-cycle fatigue region. The resulting S–N curve, after using the static strength data derived under a loading rate similar to that of the fatigue experiments, designated HLR, is similar to the Lin-Log curve (shown in Fig. 7.28) for low cycles and becomes less conservative than the Lin-Log for longer fatigue lives. On the other hand, the S–N curve derived using the static strength data obtained from standard loading rate experiments is still low, although improved, compared to the S–N curved derived without the use of any static strength values. These differences are reflected in the life prediction results presented in Fig. 7.30 where, as shown, life prediction accuracy is very much improved when the HLR static strength data are also used. In this case, the predicted results are similar to those produced using the Log-Log S–N curve.

All in all, it has been shown in this section that the examined S–N curve formulations can, under certain conditions, be used in the framework of the classic fatigue life prediction methodology and provide relatively accurate results. The Log-Log formulation based on simple linear regression analysis and easily applied to any material fatigue dataset, even by hand calculations, was proved the most reliable and provided the most accurate results within the framework of the life

Fig. 7.30 Life prediction results using S–N curves derived by the wear-out model with and without static strength data. MWX spectrum applied on-axis

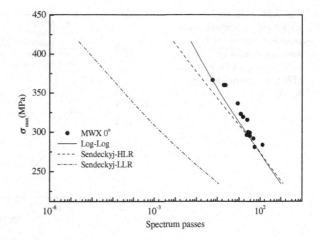

prediction methodology. However, when greater reliability is necessary, more complex models like Whitney's pooling scheme and Sendeckyj's wear-out model must be employed. In that case however, the results must be considered with caution since the derived curves can be misleading if the appropriate dataset is not analyzed, as was proved in the examined case for the wear-out model without the use of static strength data.

7.3.3.3 Effect of Constant Life Diagram Formulation

Modeling of the constant amplitude fatigue behavior of the examined material is very important within the framework of the fatigue life prediction methodology used here. One step in constant amplitude fatigue modeling is the derivation of the appropriate S–N curves, while the other is the construction of the constant life diagrams, based on which new S–N curves for any desired R-ratio must be predicted. All the models for the derivation of a constant life diagram described in Chap. 4 have been applied and selected results are presented in Fig. 7.31 for discussion. For application of the linear model, the S–N curve corresponding to reversed fatigue ($R = -1$) was used, while three S–N curves (under $R = 10$, -1 and 0.1) were used for derivation of the rest of the diagrams. Comparison of the results shows that the piecewise linear and Boerstra models are the most appropriate for the examined material under the given loading conditions and these results are presented in Fig. 7.31. The other diagrams, Harris, Harris-power, Kawai and piecewise non-linear (see Chap. 4 for details), yielded similar life predictions—those of piecewise non-linear are plotted in Fig. 7.31—while the linear diagram yielded very conservative predictions.

The explanation concerning the results presented in Fig. 7.31 is given in Fig. 7.32 where different constant life diagrams are shown for selected numbers of cycles together with the experimentally derived data from on-axis specimens

Fig. 7.31 Life prediction
results using different CLD
models. Configuration:
Rainflow-counting, Log-Log
S–N curves, CLD, Miner.
MWX spectrum applied
on-axis

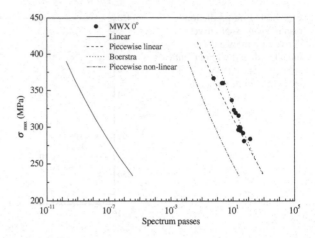

loaded under $R = 0.5$. The numbers of cycles correspond to the mean value of the estimated fatigue life per examined stress level and the relationship of the corresponding curves with the data points defines the accuracy of each presented model. As shown in this figure, the linear diagram predicts a very conservative S–N curve for the $R = 0.5$ loading case, while the piecewise non-linear yields a conservative curve but still within the experimental data range. The piecewise linear and Boerstra diagrams on the other hand are non-conservative for low numbers of cycles up to around 5×10^3 and become very accurate for numbers above 10^5. This performance is very obvious in the case of the Boerstra diagram results shown in Fig. 7.31. Use of this diagram overestimates the fatigue life of the examined material under the given loading conditions for high stress levels—low-cycle fatigue—but it becomes very accurate for lower applied loads, corresponding to higher numbers of cycles.

7.3.3.4 Effect of the Multiaxial Fatigue Failure Criterion

All multiaxial fatigue theories, except those presented by Kawai et al. and Shokrieh and Taheri (see Chap. 6 for details), were used for the prediction of the fatigue life of the examined material system under the multiaxial stress states developed as a result of the application of MWX and EPET573 variable amplitude spectra. The results, presented in Figs. 7.33, 7.34, show that all employed multiaxial fatigue failure criteria are able to produce relatively accurate life predictions, independent of the applied variable amplitude spectrum.

For application of the FWE criterion, reference curves from specimens tested under the same R-ratio as the examined one were used, although the stress ratio effect is taken into account by the criterion formulation. However, it has been proved elsewhere [26] that the selection of a reference curve from another dataset (from tests under a different stress ratio than that of the curve to be predicted) can produce highly inaccurate results. For the results shown in Figs. 7.33, 7.34 the

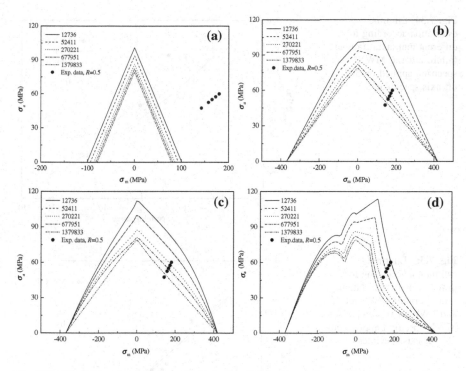

Fig. 7.32 Modeling of constant amplitude fatigue life for $R = 0.5$, on-axis specimens. **a** Linear. **b** Piecewise linear. **c** Boerstra. **d** Piecewise non-linear

Fig. 7.33 Lifetime estimation according to different multiaxial fatigue failure criteria. MWX spectrum applied on 30° off-axis specimens

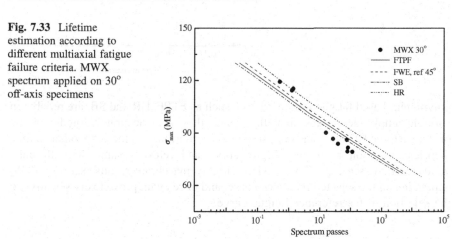

S–N curve at 45° was used as the reference for the FWE criterion, although according to the theory there is no restriction as to which curve must be used. However, the ability of the examined fatigue theories to predict fatigue life under variable amplitude complex stress states is strongly affected by their accuracy when off-axis S–N curves are estimated. Therefore, as described in Chap. 6, the

Fig. 7.34 Lifetime estimation according to different multiaxial fatigue failure criteria. MWX spectrum applied on 30° off-axis specimens

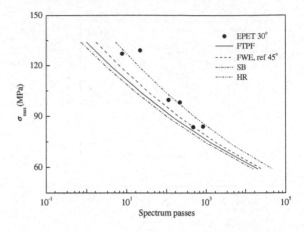

Fig. 7.35 Lifetime estimation according to different multiaxial fatigue failure criteria. FWE ref 45° and FWE ref 0° comparison. MWX spectrum applied on 30° off-axis specimens

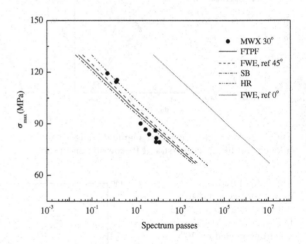

polynomial-based fatigue failure criteria, such as FTPF, HR and SB, are reliable in variable amplitude fatigue modeling since they also accurately predicted the off-axis fatigue behavior. In contrast the FWE criterion, together with other fatigue theories based on a master fatigue curve and relevant static strength data, yield accurate (see Figs. 7.33, 7.34) or highly inaccurate results (see Fig. 7.35) depending on the selected reference curve, and once again proved very sensitive to the selection of the reference fatigue curve.

7.4 Conclusions

The accuracy of the prediction of the lifetime of composite materials under variable amplitude loading is based on the accuracy of a series of processes before reaching the final goal. The theoretical background along with a thorough experimental verification of a life prediction methodology for multidirectional composite laminates under irregular load sequences has been presented in this chapter. Although the procedure could be applicable for 3D stress states, the actual implementation of both the theoretical formulation and material database is limited to plane stress conditions. Furthermore, loading spectra are assumed to produce a plane stress state in which components of the stress tensor are proportional time series. This requirement is imposed by the cycle-counting method implemented in the computational procedure.

Material property evaluation follows a direct characterization approach; the laminated plate is tested in its symmetry directions and respective fatigue strength parameters are derived. The proposed procedure is therefore suitable for the life prediction of a structural component designed for ultimate (static) or stability critical loads. If fatigue design optimization is required, a "ply-to-laminate" prediction methodology and associated ply characterization techniques should be adopted instead.

The algorithm consists mainly of five routines: cycle counting for the variable amplitude spectrum, derivation of the appropriate S–N curves for the modeling of the constant amplitude fatigue behavior of the materials, formulation of constant life diagrams to take into account the mean stress effect on the fatigue life of the examined material, derivation of allowable numbers of cycles using multiaxial fatigue failure criteria and finally, damage accumulation. Although specific choices for each one of the above routines were made, the method could in principle be implemented for any possible configuration, as the only success criterion is good prediction of the experimental results.

Nevertheless, for the cycle counting, the rainflow-counting algorithm was selected, the Log-Log S–N curve type was used together with the piecewise linear constant life diagram for the modeling of the constant amplitude fatigue data, whereas fatigue strength under multidimensional stress states was assessed using the FTPF multiaxial fatigue failure criterion because of its efficiency in satisfactorily predicting the fatigue behavior of many different composite material systems under various loading conditions. Finally, the Miner linear damage rule was adopted for the damage summation because of its simplicity and because it is the only widely accepted predictive scheme that does not require any fitting procedures on the experimental results. Furthermore, it yields successful results for GRP laminated composites in general. If necessary, non-linear regression models could be evaluated to improve the accuracy of theoretical calculations.

To demonstrate the procedure, experimental results from an extensive experimental program consisting of static and fatigue tests on specimens cut on- and off-axis from a multidirectional GRP laminate were used. Variable amplitude and constant amplitude fatigue experiments under various R-ratios were performed. Plane stress states were simulated by loading the specimens off-axis, thus developing proportional time series of stress tensor components, though of different maximum, minimum and mean values.

Implementation of the method to predict the number of spectrum passes applied to test specimens loaded on- and off-axis at 30° and 60°, using two different irregular load histories, yielded satisfactory results and proved the suitability of the procedure for the GRP laminate investigated. However, prediction accuracy was found to be dependent on several parameters such as the load spectrum, baseline data, constant life diagram, fatigue failure criterion and damage accumulation rule and this point merits further examination.

Different life predictions were also produced when static strengths, UTS and UCS, determined according to relevant standards, were used instead of those at a similar strain rate to that of the fatigue tests. Generally, it was found that if static strength data should be combined with available fatigue data in order to derive S–N curves and constant life diagrams, the static and fatigue data must be derived under the same conditions–in this case loading rates.

The use of a non-linear equation to fit the fatigue behavior of the GRP laminate investigated under variable amplitude loading was proved feasible. The resulting predictions were accurate, improving in all the examined cases those derived using the Miner linear damage rule. The major drawback in using such a non-linear life prediction scheme is the requirement of a small number of experimental data under each examined spectrum. It is therefore not practical for applications involving a large number of different fatigue load cases.

A parametric study was performed in this chapter to examine the effect of the existing solvers for the each step of the classic life prediction methodology on the theoretical results. The accuracy of some of the examined configurations in predicting the fatigue life of the examined composite material systems under both applied variable amplitude spectra is acceptable. The application of the methodology proved to be easy and straightforward without the need for any complicated numerical solutions. However, experience showed that some of the solvers can be accurate for one material and examined loading case but highly inaccurate for another, as was proved for example in this chapter with the wear-out model and constant amplitude fatigue life modeling by including and excluding static strength data.

A commonly accepted complete fatigue theory does not yet exist, although some of the sub-problems have already been successfully addressed. Software frameworks like *CCfatigue* can assist the development process of a unified fatigue life prediction methodology by comparing the life prediction results produced by different solver configurations and validating theoretical life prediction results against variable amplitude loading experimental data from different material systems.

References

1. G.P. Sendeckyj, in *Life Prediction for Resin-Matrix Composite Materials in Fatigue of Composite Materials*, vol. 4, ed. by K.L. Reifsnider, Composite Materials Series 4 (Elsevier, Amsterdam, 1991)
2. T. Adam, N. Gathercole, H. Reiter, B. Harris, Life prediction for fatigue of T800/5245 carbon-fibre composites: II Variable-amplitude loading. Fatigue **16**(8), 533–547 (1994)
3. I.P. Bond, Fatigue life prediction for GRP subjected to variable amplitude loading. Compos. Part A Appl. Sci. **30**(8), 961–970 (1999)

4. W. Hwang, K.S. Han, Cumulative damage models and multi-stress fatigue prediction. J. Compos. Mater. **20**(2), 125–153 (1986)
5. M.J. Owen, R.J. Howe, The accumulation of damage in a glass-reinforced plastic under tensile and fatigue loading. J. Phys. D Appl. Phys. **5**(9), 1637–1649 (1972)
6. J.R. Schaff, B.D. Davidson, Life prediction methodology for composite structures. Part II spectrum fatigue. J. Compos. Mater. **31**(2), 158–181 (1997)
7. L.J. Broutman, S. Sahu, A new theory to predict cumulative fatigue damage in fiberglass reinforced plastics. in *Proceedings of the 2nd Conference on Composite Materials: Testing and Design, ASTM STP 497*, (1972) pp. 170–188
8. W.X. Yao, N. Himmel, A new cumulative fatigue damage model for fibre-reinforced plastics. Compos. Sci. Technol. **60**(1), 59–64 (2000)
9. W. Hwang, K.S. Han, Fatigue of composites-fatigue modulus concept and life prediction. J. Compos. Mater. **20**(2), 154–165 (1986)
10. L.J. Lee, K.E. Fu, J.N. Yang, Prediction of fatigue damage and life for composite laminates under service loading spectra. Compos. Sci. Technol. **56**(6), 635–648 (1996)
11. W.F. Wu, L.J. Lee, S.T. Choi, A study of fatigue damage and fatigue life of composite laminates. J. Compos. Mater. **30**(1), 123–137 (1996)
12. R.P.L. Nijssen, OptiDAT–fatigue of wind turbine materials database (2006) http://www.kc-wmc.nl/optimat_blades/index.htm
13. J.F. Mandell, D.D. Samborsky, DOE/MSU Composite Material Fatigue Database. Sandia National Laboratories, SAND97-3002, (online via www.sandia.gov/wind, v. 18, 21st March (2008) Updated)
14. C.W. Kensche, in *GFRP Fatigue Data for Certification in European Wind Energy Conference Proceedings*, vol. I, (Thessaloniki, Greece, 1994) pp. 738–742
15. P.W. Bach, P.A. Joosse, D.R.V. van Delft, Fatigue lifetime of glass.polyester laminates for wind turbines. in *The European Wind Energy Conference Proceedings*, vol. I (Thessaloniki, Greece 1994), pp. 94–99
16. S.I. Andersen, P.W. Bach, W.J.A. Bonee, C.W. Kensche, H. Lilholt, A. Lystrup, W. Sys, in *Fatigue of Materials and Components for Wind Turbine Rotor Blades*. ed. by C.W. Kensche, Directorate-General XII, Science, Research and Development, EUR 16684 EN (1996)
17. T.P. Philippidis, A.P. Vassilopoulos, Life prediction methodology for GFRP laminates under spectrum loading. Compos. Part A Appl. S. **35**(6), 657–666 (2004)
18. Y. Zhang, A.P. Vassilopoulos, T. Keller, Environmental effects on fatigue behavior of adhesively–bonded pultruded structural joints. Compos. Sci. Technol. **69**(7–8), 1022–1028 (2009)
19. T.P. Philippidis, E.N. Eliopoulos, A progressive damage mechanics algorithm for life prediction of composite materials under cyclic complex stress. in *Fatigue Life Prediction of Composites and Composite Structures*, ed. by A.P. Vassilopoulos (Woodhead publishing Ltd, Cambridge 2010)
20. T.P. Philippidis, A.P. Vassilopoulos, Complex stress state effect on fatigue life of GFRP laminates. Part I, Experimental. Int. J. Fatigue **24**(8), 813–823 (2002)
21. T.P. Philippidis, A.P. Vassilopoulos, Complex stress state effect on fatigue life of GRP laminates. Part II, Theoretical formulation. Int. J. Fatigue **24**(8), 825–830 (2002)
22. T.P. Philippidis, A.P. Vassilopoulos, Fatigue strength of composites under variable plane stress. in *Fatigue in Composite*, Chap. 18. ed. by B. Harris (Woodhead Publishing and CRC Press, Cambridge 2003), pp. 504–525
23. T.W. Coats, C.E. Harris, Experimental verification of a progressive damage model for IM7/5260 laminates subjected to tension–tension fatigue. J. Compos. Mater. **29**(3), 280–305 (1995)
24. M.M. Shokrieh, L.B. Lessard, Progressive fatigue damage modelling of composite materials, Part I: Modeling. J. Compos. Mater. **34**(13), 1056–1080 (2000)
25. http://www.fireholetech.com/products/helius-fatigue.aspx
26. A.P. Vassilopoulos, R. Sarfaraz, B.D. Manshadi, T. Keller, A computational tool for the life prediction of GFRP laminates under irregular complex stress states: Influence of the fatigue failure criterion. Comp. Mater. Sci. **49**(3), 483–491 (2010)

27. ASTM E1049–85 (2005) Standard practices for cycle counting in fatigue analysis
28. R.P.L. Nijssen, Fatigue life prediction and strength degradation of wind turbine rotor blade composites. PhD Thesis, TU Delft; 2006, ISBN–13:978–90–9021221–0
29. T. Endo, K. Mitsunaga, H. Nakagawa, Fatigue of metals subjected to varying stress – Prediction of fatigue lives, preliminary proceedings of the Chigoku–Shikoku District Meeting, Japan Society of Mechanical Engineers, November 1967 pp. 41–44
30. M. Matsuishi, T. Endo, Fatigue of metals subjected to varying stress – Fatigue lives under random loading, preliminary proceedings of the Kyushu District Meeting, Japan Society of Mechanical Engineers (1968) pp. 37–40
31. J.B. de Jonge, The analysis of load–time histories by means of counting methods. in *Helicopter fatigue design guide*, ed. by F. Liard AGARD–AG–292, (November 1983)
32. S.D. Dowling, D.F. Socie, Simple rainflow counting algorithms. Int. J. Fatigue **4**(1), 31–40 (1982)
33. A. Fatemi, L. Yang, Cumulative fatigue damage and life prediction theories: A survey of the state of the art for homogeneous materials. Int. J. Fatigue **20**(1), 9–34 (1998)
34. I.P. Bond, I.R. Farrow, Fatigue life prediction under complex loading for XAS/914 CFRP incorporating a mechanical fastener. Int. J. Fatigue **22**(8), 633–644 (2000)

Index

A
3D stress states, 227
Accelerated testing, 18
Acoustic emission, 18, 50
Adaptive Neuro-Fuzzy, 89
Adherends, 138, 140, 143
Adhesively-bonded double-lap joints, 141
Advanced engineering structures, 2
Aeroelastic, 26
Aeroelastic simulation, 44, 45
Aerospace industry, 4, 11, 140
Aggressive environments, 4, 139
Aluminum tabs, 26
ANFIS, 89
Angle-ply, 11, 205
Anisomorphic, 102, 109
Anisomorphic constant fatigue
    life, 108
Anisotropic, 2, 23, 153–154, 191, 198
Antibuckling, 12, 29, 57, 60
Architectural functions, 3, 5
Artificial intelligence methods, 88
Artificial neural networks, 18, 86, 88
August wöhler, 8, 13
Automotive, 2, 140

B
Basquin relationship, 14–15
Bell-shaped, 102
Benchmarking, 38
Bi-directional, 127
Block loading, 18, 38, 193, 197, 204, 214
Boerstra's CLD, 110
Bolted joints, 3, 11, 139
Bolting, 139

Boundary conditions, 114, 115
Bridge structures, 6
Bridge widening, 6
Brittle, 2, 7, 142, 153
Buckling, 12, 57
Building applications, 2

C
Catastrophic failure, 9
CCfatigue, 199, 206, 210–212, 228
Ceramic matrix, 7
Characteristic number of cycles, 41–42,
    68–69, 72
Civil engineering, 2, 128, 141
Classic fatigue design
    methodology, 199–200, 206
Classical lamination theory, 155
CLD, 15, 17, 38, 86, 101, 103, 116, 122, 124,
    206, 216
CLD annotation, 103
CLT, 155
Coefficient of variation, 68
Complex roof shape, 6
Complex stress states, 24, 29, 155–156, 161,
    198, 225
Component investigation, 11
Composite sandwich panels, 5
Compression-compression, 29, 56–57, 86,
    103, 179
Computational methods, 102, 140
Concrete, 2, 3, 7
Confidence bounds, 56
Constant amplitude, 9, 11, 14, 26, 29, 51,
    68, 81, 85, 87, 100, 120, 156, 192,
    199, 222

A. P. Vassilopoulos and T. Keller, *Fatigue of Fiber-reinforced Composites*, 233
Engineering Materials and Processes, DOI: 10.1007/978-1-84996-181-3,
© Springer-Verlag London Limited 2011